# IMF LENDING TO
# DEVELOPING COUNTRIES

The International Monetary Fund was created to centralise the management of the global monetary system. As international financial markets evolved the richer countries turned to other more flexible sources of finance, and IMF lending became almost exclusively focused on the developing world.

And yet the IMF has been widely criticised for its lending role in developing countries, with some arguing that it should not be lending at all and others claiming that net reverse flows since the mid-1980s suggest that the Fund has abrogated its responsibilities.

This book provides the first detailed theoretical and empirical analysis of Fund lending and concludes that key changes are needed if the Fund is to realise its full potential for assisting developing countries.

**Graham Bird** is Professor of Economics and Director of the Surrey Centre for International Economic Studies, University of Surrey.

# DEVELOPMENT POLICY STUDIES
### Edited by John Farrington and Tony Killick
*for the Overseas Development Institute, London*

This series presents the results of ODI research on policy issues confronted by the governments of developing countries and their partners in aid and trade. It will be of interest to policy makers and practitioners in government and international organisations and to students and researchers in both North and South.

*Also in the series:*

## MANAGING WATER AS AN ECONOMIC RESOURCE
*James Winpenny*

# IMF LENDING TO DEVELOPING COUNTRIES

## Issues and Evidence

*Graham Bird*

London and New York

First published 1995
by Routledge
11 New Fetter Lane, London EC4P 4EE

Simultaneously published in the USA and Canada
by Routledge
29 West 35th Street, New York, NY 10001

Typeset in Garamond by
J&L Composition Ltd, Filey, North Yorkshire
Printed and bound in Great Britain by
Biddles Ltd, Guildford and King's Lynn

*British Library Cataloguing in Publication Data*
A catalogue record for this book is available from the British Library

*Library of Congress Cataloging in Publication Data*
Applied for

ISBN 0–415–11700–3

# CONTENTS

# CONTENTS

# LIST OF FIGURES

# LIST OF TABLES

# PREFACE

This book is based on a research project examining the IMF in the 1990s undertaken by the author in collaboration with Tony Killick at the Overseas Development Institute, London. Focusing on the Fund as a lending institution, it may usefully be read in conjunction with Killick's volume in the same series, *The Design and Effect of IMF Programmes in Developing Countries*, which concentrates on the Fund's role in macroeconomic stabilisation and adjustment.

At a time when there is considerable discussion as to whether the Bretton Woods institutions, as originally conceived some fifty years ago, are any longer appropriate, it is hoped that this book will make a timely, significant contribution. Within the ongoing debate, the Fund's financing role has so far been rather ignored, with much greater attention being paid to IMF conditionality and the macroeconomic consequences of Fund involvement in borrowing countries' balance of payments difficulties. However, the design of adjustment programmes cannot be divorced from the availability of finance, and the issues raised and discussed here are therefore of fundamental importance in what is a central question of international public policy.

Research also depends on finance and in this regard thanks are due to the Overseas Development Administration for financing key elements of this particular project. Many individuals have made useful comments on early drafts, including participants at a range of seminars in both the US and the UK at which parts of the research have been presented. A particular thank you must, however, go to Tony Killick for the painstaking way in which he worked through all the initial drafts. Fortunately the conclusions we reached independently are essentially consistent.

Thanks must also go to David Bedford who helped to do the

econometric work on which the Appendix to Chapter 3 is based and to Liz Blakeway, who made a huge contribution in getting the manuscript into a publishable state.

Although the project generated its own intellectual fascination, life's fun largely comes from my children, Alan, Anne, Simon and Tom, and I am grateful to them for not allowing me to become preoccupied with work. Watching them live comfortably, and happily, in a relatively rich industrialised economy serves to remind me of the practical and pressing need to ensure that international public policy does all it can to facilitate economic development in countries where children may have less to look forward to.

# 1

# THE IMF AND DEVELOPING COUNTRIES

The International Monetary Fund was originally established in order to encourage international co-operation to cope with recession and protectionism on a world scale and to discourage individual countries from pursuing policies that would beggar their neighbours and eventually themselves. The desire to improve on the international chaos of the 1930s led to the Bretton Woods Conference in 1944 and an attempt to devise a financial system which would provide a more permanent and acceptable framework for international transactions. It was intended that the emerging Bretton Woods system would generate benefits for international trade in the form of stable (though not necessarily fixed) exchange rates, whilst, at the same time, avoiding the deflationary rigidities of the gold standard mechanism. The system was designed to ensure a world of full employment and economic growth.

If the *general* purpose of the Fund at its inception was to oversee the operation of the infant Bretton Woods system, its more *specific* purposes were spelt out in Article 1 of its Articles of Agreement as follows:

(i)  To promote international monetary co-operation through a permanent institution which provides the machinery for consultation and collaboration on international monetary problems.

(ii)  To facilitate the expansion and balanced growth of international trade, and to contribute thereby to the promotion and maintenance of high levels of employment and real income and to the development of the productive resources of all members as primary objectives of economic policy.

(iii) To promote exchange stability, to maintain orderly exchange arrangements among members, and to avoid competitive exchange depreciation.

(iv) To assist in the establishment of a multilateral system of payments in respect of current transactions between members and in the elimination of foreign exchange restrictions which hamper the growth of world trade.

(v) To give confidence to members by making the general resources of the Fund temporarily available to them under adequate safeguards, thus providing them with the opportunity to correct maladjustments in their balance of payments without resorting to measures destructive of national or international prosperity.

(vi) In accordance with the above, to shorten the duration and lessen the degree of disequilibrium in the international balances of payments of members.

Within the framework set by these terms of reference the Fund operated, first, as a balance of payments adjustment institution, encouraging payments correction by means other than the use of exchange rates (except in cases of fundamental disequilibrium) or protectionist trade measures; second, as a balance of payments financing institution, providing temporary finance designed to support adjustment measures to cushion self-reversing payments instabilities; and third, as a focus for a system of rule-based international macroeconomic policy co-ordination, based essentially on the defence of established currency par values. The Fund thereby provided a linchpin for the centralised management of the international monetary system.

Throughout the 1950s and 1960s few questions were asked about the legitimacy of what the Fund was doing. On most criteria the world economy was performing well, or at least satisfactorily. Given these circumstances, the question of the extent to which success was due to the operation of the Bretton Woods system and the IMF simply did not arise. Deeper thought would have revealed the fundamental difficulties that exist in evaluating international monetary systems, and would have suggested the possibility that it was the success of the world economy which concealed the weaknesses of the Bretton Woods system and enabled it to survive. But such issues hardly seemed relevant at the time. Super-ficially at least the Fund appeared to be successful in achieving the objectives it had been set. The Bretton Woods exchange rate regime did provide a code for the non-aggressive use of devalua-tion; the IMF did provide a consultative forum within which international financial reform was debated and implemented; the world economy did enjoy a period of sustained expansion; and world trade was liberalised and did grow. So what went wrong?[1]

A lengthy answer to this question would articulate the various deficiencies of the Bretton Woods system in terms of the adjust-ment mechanism it incorporated, the method of reserve creation it used, and its vulnerability to speculative attacks. However, shorter answers are available.

The first is that the Bretton Woods system ultimately broke down. As a result, the IMF as the agency that had been charged with overseeing the operation of that system, was left disoriented. It was apparently left with little or no systemic role. The second is that the Bretton Woods system was replaced by a much looser set of international monetary arrangements. There was very little 'system' left, and, in effect, the international monetary system was privatised, with the result that there was no clear-cut role for a quasi-governmental institution such as the Fund.

Moreover, the way in which international financial arrangements evolved during the 1970s and 1980s served to marginalise the systemic role of the Fund. First, there was the adoption of general-ised flexible exchange rates. Initially these were inconsistent with the Fund's own Articles of Agreement; but even after the Articles had been modified to accommodate floating, the Fund continued to exert little influence over the direction and size of exchange rate movements. Efforts to provide a degree of surveillance over them, and to devise a set of guidelines according to which government intervention in foreign exchange markets might be carried out, had little discernible impact.

The shift to generalised flexible exchange rates also took away the means by which macroeconomic policy had been inter-nationally co-ordinated. The retreat from international policy co-ordination essentially carried on until the mid-1980s. The Bonn Summit of 1978, involving the world's leading industrial nations, represented an exception to this trend, although even here co-ordination was not organised around exchange rates. The early 1980s were characterised by the sustained appreciation in the value of the US dollar and by the 'benign neglect' that the US Administration showed for this. It was not until the Plaza Accord of 1985 that a co-ordinated move to bring down the dollar's value was implemented, with this arrangement being followed by further attempts to manage exchange rates, the most notorious of which was the Louvre Accord of 1987. However, such co-ordination was handled outside the IMF by the G-7 or the G-3 countries; the Fund's Managing Director was not even involved in the Plaza discussions. The late 1980s illustrated the degree of overlap between the former systemic role of the IMF in terms of exchange rate management, balance of payments adjustment, and the avoidance of world-wide inflation or deflation, and the actual role being contemporaneously played by a small sub-group of powerful industrial countries outside the auspices of the Fund.

The second area in which the activities of the Fund were marginalised after the breakdown in the Bretton Woods system involved balance of payments financing. Although the oil price rise of 1973 and the related acceleration and diversity in rates of inflation dramatically increased the need for international financial intermediation, this largely took place through the private international banks. The late 1970s saw the privatisation of balance of payments financing. The privatisation was, of course, neither instantaneous nor complete. The Fund did respond to the oil price rise by introducing the Oil Facility and it did expand its loans in the mid-1970s by liberalising the Compensatory Financing Facility (CFF). But the extent of privatisation was on a sufficient scale to reflect a broad systemic change. During the period 1977–82 short-term bank lending to developing countries ran at an annual average of \$19.5 bn, whereas net purchases from the Fund were only \$2.5 bn.

Related to both the move to exchange rate flexibility and the private financing of balance of payments deficits, the Fund's importance as a source of official reserve creation also became marginalised during the 1970s. Ironically, the decade had begun with the introduction of the Special Drawing Right; and even as late as 1976, at its Jamaica meetings, the Fund was setting the objective of establishing the SDR as the principal reserve asset in the international financial system. The reality, however, was that, with flexible exchange rates and the private financing of payments deficits, the quantity of official reserves became viewed as an unimportant issue. The 'system' moved over to the wider use of certain national currencies as international reserve assets, thereby becoming a multiple reserve currency system; SDR creation was not maintained, and attempts to introduce a substitution account failed; even the SDR's use as a unit of account was superseded by the European Community's European Currency Unit (ECU). Some critics observed gleefully that, rather than just being marginalised, the SDR had been almost obliterated. Certainly no effective role seemed to be left for the Fund in influencing global reserve adequacy – a role that had appeared central in the 1960s.[2]

Finally, the trend seemed to be to move away from international monetary arrangements and towards regional ones. Increasing regionalisation was most dramatically illustrated by the establishment of the European Monetary System in 1979. It was now at the regional level that the management of exchange rates and the co-ordination of macroeconomic policy occurred.

While such developments led some people to call for the establishment of a European Monetary Fund to carry out within Europe the former systemic functions of the IMF, others now argued for the dissolution of the IMF. During the 1970s this call was loudest from those developing countries which, observing the collapse of the old economic order, of which the Fund was seen as a central part, urged the establishment of a New International Economic Order (NIEO) with brand-new institutions. However, the political influence of this argument was only ever likely to be as strong as the commodity cartels that in fact failed to materialise.[3] More influential remained the argument that the Fund was no longer needed in a non-Bretton Woods and market-dominated world economy. What was there left for the Fund to do?

## CHANGING PARTNERS: THE FUND'S INVOLVEMENT WITH DEVELOPING COUNTRIES

While events during the 1970s undermined the global systemic role of the Fund, they also served to create circumstances in which it was almost forced into accepting a new and more specific role.[4] This role reflected the evolving balance of payments problems which developing countries encountered during the 1970s and 1980s. During the latter part of the 1970s the Fund's role was largely limited to the world's poorest countries, located in Africa and Asia, which lacked creditworthiness in the eyes of the commercial banks. Beyond 1982, however, and with the arrival of the

Table 1.1 Developing countries: net credit from IMF, 1982–92 (US$ bn)[a]

| Developing countries | 1982 | 1983 | 1984 | 1985 | 1986 | 1987 | 1988 | 1989 | 1990 | 1991 | 1992 |
|---|---|---|---|---|---|---|---|---|---|---|---|
|  | 6.9 | 11.0 | 4.7 | 0.3 | -2.2 | -4.7 | -4.1 | -1.5 | -1.9 | 1.1 | -0.2 |
| *By region* | | | | | | | | | | | |
| Africa | 2.0 | 1.3 | 0.6 | 0.1 | -1.0 | -1.1 | -0.3 | 0.1 | -0.6 | 0.2 | -0.2 |
| Asia | 2.3 | 2.5 | 0.3 | -1.0 | -0.9 | -2.4 | -2.4 | -1.1 | -2.4 | 1.9 | 1.3 |
| Middle East & Europe | 1.2 | 1.1 | 0.5 | -0.2 | -0.5 | -0.4 | -0.5 | -0.2 | -0.1 | – | 0.4 |
| Western Hemisphere | 1.5 | 6.1 | 3.4 | 1.5 | 0.1 | -0.8 | -0.9 | -0.2 | 1.2 | -1.0 | -1.6 |
| Sub-Saharan Africa | 0.7 | 1.3 | 0.5 | – | -0.4 | -0.5 | -0.2 | -0.4 | -0.3 | – | – |
| *By predominant export* | | | | | | | | | | | |
| Fuel | 0.2 | 1.7 | 1.3 | – | 0.8 | 1.0 | – | 2.0 | 2.7 | 0.3 | -1.3 |
| Non-fuel exports | 6.7 | 9.3 | 3.5 | 0.3 | -3.0 | -5.7 | -4.1 | -3.5 | -4.6 | 0.8 | 1.2 |
| Manufactures | 3.6 | 4.7 | 3.0 | -0.6 | -1.0 | -4.0 | -2.9 | -2.6 | -2.6 | 1.3 | 1.1 |
| Primary products | 1.2 | 3.4 | 0.6 | 1.1 | -0.5 | -0.3 | -0.4 | -1.0 | -0.9 | -0.8 | -0.4 |
| Services and private transfers | 0.6 | 0.6 | – | -0.2 | -0.6 | -0.6 | -0.6 | 0.2 | -0.4 | 0.3 | 0.3 |
| Diversified export base | 1.5 | 0.5 | – | – | -1.0 | -0.8 | -0.3 | -0.1 | -0.7 | 0.1 | 0.1 |

Source: IMF, *World Economic Outlook*, Washington, DC, October 1990 and May 1993.

Note. [a] Includes net disbursements from programmes under the General Resources Account Trust Fund, structural adjustment facility (SAF) and enhanced structural adjustment facility (ESAF). Projected net disbursements incorporate the impact of prospective programmes. The data are on a transactions flow basis, with conversions to US dollar values at annual average exchange rates.

*By financial criteria*

| | | | | | | | | | | | |
|---|---|---|---|---|---|---|---|---|---|---|---|
| Net creditor countries | — | — | — | — | — | — | — | — | — | — | — |
| Net debtor countries | 6.9 | 11.0 | 4.7 | 0.3 | -2.2 | -4.7 | -4.1 | -1.5 | -1.9 | 1.1 | -0.2 |
| Market borrowers | 2.0 | 6.1 | 3.8 | 1.2 | 0.6 | -1.8 | -1.4 | 0.2 | 0.7 | -1.2 | -1.6 |
| Diversified borrowers | 3.1 | 3.3 | 0.6 | — | — | — | — | — | — | — | — |
| Official borrowers | 1.8 | 1.6 | 0.3 | -0.2 | -1.0 | -0.8 | -0.8 | -0.2 | -1.1 | 0.3 | 0.3 |
| Countries with recent debt-servicing difficulties | 4.0 | 7.9 | 3.9 | 1.7 | -1.0 | -1.8 | -1.3 | -0.5 | 0.4 | -1.0 | -1.8 |
| Countries without debt-servicing difficulties | 2.9 | 3.1 | 0.9 | -1.4 | -1.2 | -2.9 | -2.8 | -1.0 | -2.3 | 2.1 | 1.6 |

*Miscellaneous groups*

| | | | | | | | | | | | |
|---|---|---|---|---|---|---|---|---|---|---|---|
| Small low-income economies | 1.0 | 1.2 | 0.2 | -0.2 | -0.9 | -0.6 | -0.3 | — | -0.6 | 0.4 | 0.2 |
| 15 heavily indebted countries | 2.2 | 6.3 | 3.3 | 1.6 | -0.2 | -1.3 | -1.4 | -0.8 | 0.6 | -1.4 | -1.8 |

debt crisis, its dealings spread to include the better-off developing countries of Latin America. The Fund's involvement included both a financing and an adjustment element, and it was in the context of the period immediately following the debt crisis that it transiently recaptured systemic significance by seeking to avert the collapse of the international banking system which some commentators argued the debt crisis would cause.

The Fund's involvement with developing countries is quite starkly revealed by considering the size and pattern of its lending since the mid-1970s.

Table 1.1 gives information on the use of Fund credit in the period 1982–92. This is confined to developing countries because no industrial country has used Fund credit in recent years. The picture it provides contrasts sharply with that for earlier periods. In 1968–72, for example, 11 industrial countries, including all the G-7 countries (with the sole exception of Japan), drew on the Fund. Drawings by developing countries, although relatively numerous, were also relatively small. Thus, over the same five-year period, 33 developing countries used Fund credit, but even at their peak in 1968 these drawings reached only 23 per cent of the total quotas of developing countries. In 1970, SDR 2.4 bn of the Fund's total outstanding credit of SDR 3.2 bn was with industrial countries. Throughout the rest of the 1970s, industrial countries remained substantial users of Fund credit in one form or another, in spite of the move to flexible exchange rates. Industrial countries were not insulated from the balance of payments consequences of the oil price rise of 1973 and there were some doubts concerning the permanency of flexible rates. Indeed, drawings by the United Kingdom and Italy on their own accounted for almost 40 per cent of total drawings from the Fund during 1974–7. Beyond this period, however, no major (G-10) industrial country has used Fund credit. The United States drew under the reserve tranche in November 1978, but this did not involve the use of Fund credit. Indeed, significantly there is some evidence that this drawing, as well as an earlier drawing by the UK in 1976, had been motivated by political considerations, basically the desire to obtain an outside endorsement for unpopular domestic policy.

With the legitimisation of floating exchange rates through the amendment of the Fund's Articles of Agreement in 1978; the establishment of the European Monetary System in 1979 and the credit facilities which this provided for members of the

system; and continuing innovation in international financial intermediation using private capital markets, industrial countries could
now easily bypass the IMF.

This was clearly not the case for developing countries, particularly after the debt crisis had come to the fore in 1982. Prior
to that date a limited number of the better-off developing countries had enjoyed access to private capital in the form of loans from
the commercial banks. By late 1981 most Fund lending was to the
least developed countries. This is reflected in Table 1.1 by the
relatively large amount of net Fund credit to the developing
countries of Africa and Asia observed in 1982 and by the relatively
small amount of net credit to the less developed countries of the
Western Hemisphere. At this time a division of labour appeared to
be emerging between the Fund and the banks in terms of lending
to developing countries. However, the pattern was rudely disturbed
in 1983 and beyond, as countries formerly deemed creditworthy by
the banks found their access to commercial credit being cut off.
Given the size of their adjustment problems, these countries were
now forced to turn to the Fund for finance. While in 1982 Fund
credit in African developing countries had been a third larger than
in those of the Western Hemisphere, by 1983 it was only a fifth of
the Western Hemisphere figure. The change in the pattern of Fund
lending to developing countries was indeed dramatic. While there
had been a steady increase in outstanding Fund credit in African
developing countries during the period 1970–85, and a rather less
steady increase in Asia, Fund credit outstanding in the developing
countries of the Western Hemisphere actually fell between 1976
and 1981 but then increased tenfold in the next five years.

Yet while the first half of the 1980s saw the Fund becoming
quite heavily involved in providing credit to developing countries,
by the second half of the decade the net transfer had become
negative. If the negative net transfer that developing countries
faced in terms of the banks was a problem, the Fund seemed to
be adding to it rather than helping to resolve it. On the other hand,
the positive net flow of Fund resources in the earlier 1980s seen
against a negative net flow of commercial loans had led to
accusations that the Fund was bailing out the banks.

In quantitative terms, the Fund's response to the changing
financing role of the banks during the 1980s was only partial.
Even when the Fund's net lending was positive in the first half
of the decade, it did not offset the negative net flows associated

*Table 1.2* Summary of external financing of balance of payments deficits in developing countries, 1985–93 (US$ bn)

| | 1985 | 1986 | 1987 | 1988 | 1989 | 1990 | 1991 | 1992 | 1993 |
|---|---|---|---|---|---|---|---|---|---|
| *Developing countries* | | | | | | | | | |
| Balance on current account, excluding official transfers | −42.4 | −64.5 | −24.5 | −43.8 | −38.1 | −33.4 | −78.6 | −96.2 | −84.5 |
| Changes in reserves (−= increase) | −11.1 | −10.0 | −50.4 | 3.5 | −25.2 | −49.9 | −66.8 | −49.8 | −26.0 |
| Non-debt-creating flows, net | 22.8 | 24.1 | 28.8 | 31.2 | 32.4 | 31.4 | 26.7 | 54.0 | 48.3 |
| Official transfers | 12.8 | 13.8 | 14.8 | 15.1 | 16.1 | 12.1 | −3.3 | 17.8 | 10.1 |
| Direct investment | 9.9 | 10.3 | 14.0 | 16.1 | 16.3 | 19.3 | 30.1 | 36.2 | 38.2 |
| Net credit from IMF | 0.3 | −2.2 | −4.7 | −4.1 | −1.5 | −1.9 | 1.1 | −0.2 | – |
| Net long-term borrowing from official creditors | 24.1 | 31.0 | 26.9 | 21.6 | 28.6 | 47.6 | 30.7 | 42.3 | 35.8 |
| Net borrowing from commercial banks | 4.9 | 1.4 | 8.2 | 2.5 | −1.7 | 26.5 | 36.1 | 5.5 | 16.2 |

*Countries is transition:*

*Central Europe*

| | | | | | | | | | |
|---|---|---|---|---|---|---|---|---|---|
| Balance on current account, excluding official transfers | 1.2 | 0.1 | 2.1 | 5.9 | 2.7 | −0.2 | −6.6 | −1.2 | −2.7 |
| Changes in reserves (−= increase) | −0.6 | −1.7 | 0.1 | −3.2 | −4.6 | 0.7 | −2.1 | −2.1 | −4.1 |
| Asset transactions, including net errors and omissions | −1.4 | 0.3 | −1.6 | 2.9 | 2.4 | −2.5 | 2.2 | −0.1 | −0.8 |
| Non-debt-creating flows, net | — | — | — | — | 0.5 | 1.0 | 2.5 | 3.8 | 2.9 |
| Official transfers | — | — | — | — | 0.1 | 0.2 | 0.1 | 0.7 | 0.4 |
| Direct investment | 0.1 | — | — | — | 0.4 | 0.8 | 2.4 | 3.1 | 2.5 |
| Reserved-related liabilities | −0.3 | −0.5 | −1.1 | −0.9 | −0.7 | −0.2 | 3.5 | 0.6 | 1.5 |
| Net credit from IMF | −0.3 | −0.5 | −1.1 | −0.9 | −0.9 | 0.1 | 3.5 | 0.6 | — |
| Net external borrowing | 1.2 | 1.9 | 0.5 | −4.6 | −0.3 | 1.2 | 0.5 | −1.0 | 3.2 |

*Source:* IMF, *World Economic Outlook*, Washington, DC, May 1993.

with the commercial banks. In any case, the response was largely forced on the Fund by outside events; it was not a response that the Fund had itself orchestrated, except to the extent that concerted lending had been seen as a means of averting a major international financial catastrophe as a consequence of the debt crisis. Basically, if countries are eligible to draw, then the Fund has to make resources available. The changing pattern of Fund lending largely reflected changes in the demands coming from developing countries rather than a desire on the part of the Fund to become more heavily involved in lending to them.

By the beginning of the 1990s data in Table 1.2 show that commercial bank lending was again relatively heavy to developing countries as a whole. Net credit from the IMF seemed small relative to all other forms of external financing.

Whatever its cause, the changing pattern of Fund lending raised a series of questions concerning the role of the Fund. Should it be lending exclusively to developing countries? Had it become, to all intents and purposes, a development agency? Was the nature of its conditionality appropriate for the countries that were now turning to it? Did lending to developing countries not mean that there was considerable overlap between the Fund and the World Bank, and if so, how should this overlap be handled? Was the exclusivity of its clientele causing the Fund to lose sight completely of its former systemic role? Was there a need to make a distinction at the very least between the problems and requirements of the middle-income as compared with the low-income countries? Underpinning much of this was a more general question which the Fund had wrestled with during much of its history – the question of whether developing countries warranted special treatment within the international monetary system.

## DEVELOPING COUNTRIES AS A SPECIAL CASE

In its early years the Fund rejected the claim that developing countries warranted any form of special treatment. However, over the years a number of reforms were implemented which were primarily or exclusively directed towards developing countries. Given the Fund's position as a balance of payments institution, the rationale for such reforms has not been that of international equity, but rather the implicit acceptance that developing countries encounter payments problems that are different in either size or nature from those encountered by other country groups. What criteria might reflect this?

Balance of payments difficulties emanate from a number of sources. First, there may be secular changes in exports, imports and long-term capital flows. For example, a country producing and exporting goods that have a low income elasticity of demand and importing goods that have a higher one will tend to encounter balance of payments problems. Such factors reflect payments deficits and surpluses as essentially structural phenomena. On top of this, where demand and supply are themselves unstable, low price elasticities of demand and supply will tend to result in instability in the terms of trade. In part such balance of payments instability reflects vulnerability to exogenous shocks.

Both the above factors influence the incidence of payments deficits and surpluses. Other important aspects of the balance of payments relate to the speed and efficiency with which deficits may be financed or corrected. The capacity of a country to finance a payments deficit depends on the level of its reserve holdings and the availability of finance from the private international banks and the Bretton Woods institutions. In turn, the scope for payments correction varies with the capacity for adjustment within the economy. This depends on a number of factors, including the extent to which domestic consumption may be switched into exports, and more generally, the scope for short-run export expansion and efficient import substitution, the degree of money illusion, the flexibility of domestic economic policy, the level of infrastructural investment and, related to the above, the values of export supply and import and export demand price elasticities.

For example, with low elasticities and a high degree of real wage resistance the scope for balance of payments adjustment will be strictly constrained. Clearly to the extent that the adaptability of an economy is positively related to its level of economic development it is likely that developing countries will encounter more difficulty in coping with balance of payments problems than do developed economies. However, the presumption may not always be valid. It is not difficult to think of developing countries that have been characterised by their ability to respond to a changing world economic environment. Similarly, one can think of developed countries that find change difficult to accommodate because of their stymied socio-economic and political systems. An important feature of the 1970s and 1980s was the growing irrelevance of a categorisation of countries that lumped all developing countries together. Such an approach may be positively misleading. Disaggregation is therefore vitally important. This may be based on various economic indicators including *per capita* income, the degree of export diversification, the nature and pattern of trade, and geographical location (see Table 1.1 again). Against this background, a number of indicators may be assembled to provide some reflection of the size and nature of the payments problem of a country or countries.

## Balance of payments trends

Table 1.3 reveals that during the period 1985–93 most developing countries, with the exception of the newly industrialising Asian economies, experienced quite persistent current account deficits. When expressed as a percentage of exports of goods and services this tended to be much greater for the low-income countries than for the better-off developing countries. Data from the IMF's *World Economic Outlook* also reveal that declines in trade, and particularly import volumes, have not been uncommon, especially again in the low-income countries. To the extent that there have been improvements in the trade balance of highly indebted developing countries these have frequently occurred at a lower level of trade.

Of course, the statistical state of the current account balance of payments is an imperfect guide to the size of payments problems. Disequilibria may be temporary and self-reversing; capital flows may allow a current account deficit to be sustained; and *ex post* payments data may conceal the extent to which other macroeconomic policy objectives have been subjugated. Moreover, experience in the 1970s and 1980s clearly demonstrates that the financing of trade and non-factor service deficits by private borrowing will be unsustainable unless it is accompanied by a large enough increase in the capacity to meet the related debt obligations.

The nature of the international economy may dictate that such capacity has to be created by means of compressing imports. The policies through which this is achieved, although statistically strengthening the balance of payments, are likely to damage aspects of domestic economic performance in the short run and quite possibly balance of payments performance in the long run.[5]

Although there are significant differences to be found between

14

Table 1.3 Developing countries: payments balances on current account, 1985–93 (US$ bn)

| | 1985 | 1986 | 1987 | 1988 | 1989 | 1990 | 1991 | 1992 | 1993 |
|---|---|---|---|---|---|---|---|---|---|
| Developing countries | -29.6 | -50.7 | -9.7 | -28.7 | -22.0 | -21.3 | -81.9 | -78.4 | -74.4 |
| *By region* | | | | | | | | | |
| Africa | -0.5 | -9.4 | -4.5 | -9.6 | -6.5 | -2.3 | -4.0 | -7.8 | -6.8 |
| Asia | -15.9 | 2.0 | 18.0 | 6.3 | -4.5 | -6.5 | -6.6 | -21.2 | -27.3 |
| Middle East & Europe | -9.1 | -24.8 | -12.0 | -13.8 | -2.5 | -5.6 | -53.0 | -15.8 | -6.9 |
| Western Hemisphere | -4.1 | -18.6 | -11.1 | -11.6 | -8.4 | -6.8 | -18.4 | -33.5 | -33.4 |
| Sub-Saharan Africa | -3.4 | -5.5 | -6.1 | -7.4 | -6.2 | -8.1 | -8.7 | -9.4 | -8.8 |
| Four newly industrialising Asian economies | 8.3 | 21.2 | 27.7 | 25.2 | 18.9 | 10.8 | 7.5 | -1.5 | -3.0 |
| *By predominant export* | | | | | | | | | |
| Fuel | -2.8 | -39.4 | -12.4 | -29.6 | -10.9 | -4.2 | -68.4 | -45.7 | -35.4 |
| Non fuel exports | -26.8 | -11.4 | 2.7 | 0.9 | -11.1 | -17.1 | -13.5 | -32.6 | -39.0 |
| Manufactures | -9.9 | 2.6 | 18.8 | 17.6 | 7.4 | 0.3 | 6.3 | -3.5 | -9.3 |
| Primary products | -10.1 | -11.1 | -14.8 | -12.5 | -11.1 | -10.5 | -13.4 | -20.2 | -19.7 |
| Agricultural products | -7.8 | -7.6 | -11.2 | -9.7 | -9.4 | -6.7 | -9.8 | -15.8 | -14.8 |
| Minerals | -2.3 | -3.5 | -3.6 | -2.7 | -1.8 | -3.8 | -3.5 | -4.4 | -4.9 |
| Services & private transfers | -6.4 | -5.1 | -5.2 | -5.9 | -6.2 | -5.5 | -3.8 | -2.6 | -2.8 |
| Diversified export base | -0.4 | 2.3 | 3.8 | 1.7 | -1.2 | -1.4 | -2.6 | -6.3 | -7.1 |
| *By financial criteria* | | | | | | | | | |
| Net creditor countries | 11.2 | 5.5 | 14.7 | 5.0 | 13.5 | 13.4 | -40.9 | -16.8 | -9.5 |
| Net debtor countries | -40.8 | -56.2 | -24.4 | -33.7 | -35.6 | -34.7 | -41.0 | -61.6 | -64.8 |
| Market borrowers | -13.4 | -20.0 | 4.1 | -0.1 | -4.9 | 0.3 | -20.1 | -35.7 | -38.3 |
| Diversified borrowers | -15.9 | -19.4 | -15.7 | -18.5 | -16.7 | -24.7 | -9.4 | -14.5 | -14.7 |

Table 1.3 (continued)

| | 1985 | 1986 | 1987 | 1988 | 1989 | 1990 | 1991 | 1992 | 1993 |
|---|---|---|---|---|---|---|---|---|---|
| *Developing countries* | −29.6 | −50.7 | −9.7 | −28.7 | −22.0 | −21.3 | −81.9 | −78.4 | −74.4 |
| Official borrowers | −11.5 | −16.8 | −12.8 | −15.1 | −13.9 | −10.4 | −11.5 | −11.4 | −11.8 |
| Countries with recent debt-servicing difficulties | −15.1 | −40.4 | −24.0 | −29.9 | −22.3 | −19.8 | −27.4 | −42.6 | −41.8 |
| Countries without recent debt-servicing difficulties | −25.7 | −15.9 | −0.4 | −3.7 | −13.2 | −14.9 | −13.5 | −18.9 | −23.0 |
| *Other groups* | | | | | | | | | |
| Small low-income incomes | −7.3 | −7.1 | −7.9 | −9.4 | −9.8 | −11.0 | −11.3 | −11.5 | −11.5 |
| Least developed countries | −4.7 | −4.3 | −4.5 | −5.2 | −4.8 | −6.4 | −6.4 | −6.4 | −6.2 |
| 15 heavily indebted countries | −1.2 | −18.9 | −9.7 | −10.6 | −6.5 | −4.3 | −21.7 | −33.8 | −33.9 |

*In % of exports of goods and services*

| | Average 1975–84 | 1985 | 1986 | 1987 | 1988 | 1989 | 1990 | 1991 | 1992 | 1993 |
|---|---|---|---|---|---|---|---|---|---|---|
| *Developing countries* | −2.3 | −4.9 | −8.8 | −1.4 | −3.6 | −2.5 | −2.1 | −7.7 | −6.8 | −5.9 |
| *By region* | | | | | | | | | | |
| Africa | −14.2 | −0.6 | −13.5 | −5.8 | −11.9 | −7.5 | −2.3 | −4.0 | −8.0 | −6.7 |
| Asia | −5.3 | −6.7 | 0.8 | 5.3 | 1.5 | −1.0 | −1.2 | −1.1 | −3.2 | −3.7 |
| Middle East & Europe | 13.2 | −5.4 | −18.7 | −8.0 | −8.7 | −1.4 | −2.6 | −25.5 | −7.3 | −3.0 |
| Western Hemisphere | −22.6 | −3.3 | −16.9 | −9.0 | −8.3 | −5.5 | −4.1 | −10.8 | −18.7 | −17.2 |
| Sub-Saharan Africa | −24.3 | −13.3 | −20.9 | −21.5 | −25.1 | −19.7 | −23.7 | −25.9 | −28.7 | −25.4 |
| Four newly industrialising Asian economies | −2.8 | 6.5 | 13.8 | 13.6 | 9.8 | 6.6 | 3.4 | 2.1 | −0.4 | −0.7 |

*By predominant export*

| | | | | | | | | | |
|---|---|---|---|---|---|---|---|---|---|
| Fuel | 9.0 | −1.2 | −23.6 | −6.5 | −15.2 | −4.7 | −1.5 | −25.7 | −16.8 | −12.2 |
| Non-fuel exports | −14.0 | −7.1 | −2.8 | 0.5 | 0.2 | −1.7 | −2.3 | −1.7 | −3.7 | −4.0 |
| Manufactures | −10.5 | −4.2 | 1.0 | 5.6 | 4.2 | 1.6 | 0.1 | 1.1 | −0.5 | −1.3 |
| Primary products | −24.3 | −19.2 | −19.9 | −25.2 | −18.7 | −15.4 | −13.1 | −16.4 | −24.1 | −21.7 |
| Agricultural products | −22.5 | −21.2 | −19.2 | −27.8 | −21.4 | −19.7 | −12.3 | −17.7 | −27.7 | 23.6 |
| Minerals | −29.4 | −14.4 | −21.7 | −19.5 | −13.0 | −7.2 | −14.9 | −13.6 | −16.5 | −17.3 |
| Services & private transfers | −19.7 | −20.1 | −15.5 | −15.0 | −15.7 | −15.4 | −12.4 | −8.0 | −5.0 | −5.0 |
| Diversified export base | −13.4 | −0.7 | 3.9 | 5.6 | 2.2 | −1.4 | −1.5 | −2.6 | −5.6 | −5.8 |

*By financial criteria*

| | | | | | | | | | |
|---|---|---|---|---|---|---|---|---|---|
| Net creditor countries | 20.4 | 7.6 | 4.3 | 9.7 | 3.0 | 7.1 | 6.2 | −18.3 | −7.2 | −3.9 |
| Net debtor countries | −2.8 | −8.9 | −12.6 | −4.5 | −5.3 | −5.1 | −4.4 | −4.8 | −6.7 | −6.3 |
| Market borrowers | −13.6 | −4.7 | −7.3 | 1.2 | – | −1.0 | 0.1 | −3.4 | −5.5 | −5.3 |
| Diversified borrowers | −11.7 | −13.3 | −16.7 | −11.8 | −12.8 | −10.5 | −14.3 | −5.4 | −7.9 | −7.1 |
| Official borrowers | −22.2 | −19.4 | −31.4 | −21.5 | −22.8 | −19.1 | −12.2 | −13.2 | −12.7 | −12.2 |
| Countries with recent debt-servicing difficulties | −19.6 | −7.3 | −22.5 | −11.8 | −13.3 | −9.0 | −7.3 | −10.3 | −15.4 | −13.8 |
| Countries without recent debt-servicing difficulties | −22.5 | −10.1 | −6.0 | −0.1 | −0.9 | −2.9 | −2.9 | −2.3 | −2.9 | −3.2 |

*Other groups*

| | | | | | | | | | |
|---|---|---|---|---|---|---|---|---|---|
| Small low-income economies | −37.0 | −33.1 | −30.3 | −31.2 | −34.0 | −34.1 | −34.2 | −34.2 | −33.4 | −30.4 |
| Least developed countries | −38.1 | −31.7 | −28.4 | −26.6 | −27.5 | −23.8 | −29.6 | −29.9 | −29.6 | −26.5 |
| 15 heavily indebted countries | −21.3 | −0.8 | −14.7 | −6.6 | −6.5 | −3.6 | −2.1 | −11.0 | −16.1 | −15.0 |

*Memorandum Median*

| | | | | | | | | | |
|---|---|---|---|---|---|---|---|---|---|
| Developing countries | −16.5 | −12.5 | −12.2 | −13.3 | −11.6 | −10.7 | −12.0 | −11.7 | −10.5 | −8.3 |

*Source:* IMF, *World Economic Outlook*, Washington, DC, May 1993.

different groupings of developing countries and different time periods, the empirical evidence generally supports the claim that developing countries have experienced relatively severe balance of payments problems by comparison with the rest of the world. This conclusion is superficially confirmed by the evidence presented earlier showing drawings on IMF resources. A prerequisite for Fund support is the existence of a balance of payments need, and the evidence suggests that recently it has been *only* in developing countries that this need has reached proportions where Fund assistance was warranted.

## Commodity terms of trade, export concentration and export stability

The prolonged downward trend in the price of primary commodities relative to manufactures has for a long time been claimed to present developing countries with particular problems. Table 1.4 provides evidence on their terms of trade and again shows the dangers of generalisation. For some developing countries which export manufactures the terms of trade, taking 1985–93 as a whole, have improved, whereas for others, such as those in sub-Saharan Africa, adverse movements have been particularly marked. For these economies what is happening to primary product prices is of particular relevance, given their degree of export concentration on such products. UNCTAD data covering 1982–4 show that for 58 per cent of all developing countries, primary products made up more than 50 per cent of their total exports, but that 71 per cent of low-income countries possessed this degree of export concentration.

Moreover, while a downward trend in the commodity terms of trade of developing countries (with no offsetting improvement in their income terms of trade) suggests that they will encounter a secular deterioration in their balance of payments, the problems of managing the related difficulties are exaggerated by short-term instability about this trend. Although the subject of extended academic debate, the extent to which export instability creates a special problem for developing countries has been recognised institutionally by the introduction within the IMF of what was the Compensatory Financing Facility, and by the European Community's Stabex Scheme. However, recent evidence suggests that, while instability declined during the 1960s, it increased sub-

Table 1.4 Developing countries: terms of trade, 1975–94 (% annual change)

| | Average 1975–84 | 1985 | 1986 | 1987 | 1988 | 1989 | 1990 | 1991 | 1992 | 1993 | 1994 |
|---|---|---|---|---|---|---|---|---|---|---|---|
| Developing countries | 2.1 | -2.1 | -16.8 | 1.9 | -4.8 | 4.1 | 2.1 | -3.8 | -1.4 | -1.0 | -0.1 |
| *By region* | | | | | | | | | | | |
| Africa | 0.4 | -0.7 | -21.1 | 0.5 | -3.9 | -0.6 | 3.0 | -5.5 | -5.5 | -2.8 | -0.8 |
| Asia | – | -0.2 | -2.9 | – | -1.8 | 3.9 | -1.4 | -0.7 | -0.4 | -0.9 | -0.3 |
| Middle East & Europe | 4.8 | -3.8 | -41.2 | 13.5 | -16.8 | 10.0 | 14.1 | -11.1 | -3.3 | -1.1 | 0.1 |
| Western Hemisphere | 0.4 | -5.1 | -9.5 | -4.7 | -0.5 | 1.2 | -0.6 | -4.0 | -0.6 | -0.2 | 0.8 |
| Sub-Saharan Africa | -1.0 | -0.3 | -8.6 | -7.4 | 1.4 | -3.1 | -4.0 | -4.3 | -4.3 | 0.4 | -0.3 |
| Four newly industrialising Asian economies | -0.4 | 2.2 | 4.1 | -1.2 | -2.0 | 7.9 | -1.6 | -0.4 | -0.4 | -1.6 | -0.6 |
| *By predominant export* | | | | | | | | | | | |
| Fuel | 4.8 | -4.8 | -46.6 | 11.5 | -16.1 | 12.0 | 15.7 | -12.4 | -3.2 | -2.3 | – |
| Non-fuel exports | -0.9 | -0.4 | 3.0 | -1.9 | -0.5 | 1.6 | -2.5 | -0.5 | -0.8 | -0.6 | -0.2 |
| Manufactures | -1.0 | 0.8 | 5.3 | -2.1 | -1.8 | 3.8 | -2.1 | – | -0.4 | -0.8 | -0.3 |
| Primary products | -0.6 | -4.5 | -2.5 | -6.6 | 4.7 | -2.7 | -6.3 | -3.0 | -1.1 | 0.1 | 0.7 |
| Agricultural products | 0.1 | -4.3 | -2.2 | -10.0 | -0.3 | -4.1 | -5.7 | -2.9 | -0.8 | 2.4 | 1.2 |
| Minerals | -2.2 | -5.3 | -3.6 | 0.6 | 14.6 | 0.1 | -7.0 | -3.2 | -1.7 | -4.7 | -0.5 |
| Services & private transfers | -0.5 | 1.8 | 1.4 | -6.9 | 4.9 | -0.5 | -2.7 | -1.0 | -3.4 | 1.9 | 2.1 |
| Diversified export base | -0.8 | -2.3 | -2.6 | 4.0 | 1.1 | -4.5 | -1.6 | -0.5 | -0.4 | -0.9 | -0.4 |

*Source:* IMF, *World Economic Outlook*, Washington, DC, May 1993.

Table 1.5 Developing countries: ratios of reserves to imports of goods and services, 1985–93[a,b]

| | 1985 | 1986 | 1987 | 1988 | 1989 | 1990 | 1991 | 1992 | 1993 |
|---|---|---|---|---|---|---|---|---|---|
| *Developing countries* | 27.5 | 29.5 | 34.2 | 28.6 | 27.9 | 29.2 | 32.3 | 33.1 | 32.3 |
| *By region* | | | | | | | | | |
| Africa | 11.4 | 10.6 | 11.9 | 10.7 | 12.0 | 15.0 | 18.4 | 18.4 | 20.1 |
| Asia | 28.3 | 36.7 | 42.8 | 36.0 | 33.5 | 34.1 | 36.7 | 35.3 | 33.0 |
| Middle East & Europe | 31.3 | 30.9 | 35.1 | 28.8 | 29.7 | 27.4 | 26.4 | 28.6 | 30.2 |
| Western Hemisphere | 31.4 | 25.4 | 27.4 | 19.8 | 19.8 | 26.2 | 33.5 | 38.7 | 38.2 |
| Sub-Saharan Africa | 12.3 | 13.9 | 14.8 | 15.1 | 16.1 | 17.5 | 19.9 | 21.8 | 24.7 |
| Four newly industrialising Asian economies | 32.1 | 47.3 | 54.3 | 44.9 | 41.4 | 38.3 | 37.4 | 34.7 | 30.9 |
| *By predominant export* | | | | | | | | | |
| Fuel | 32.7 | 29.4 | 37.0 | 24.7 | 25.2 | 26.5 | 28.2 | 26.3 | 24.9 |
| Non-fuel exports | 24.9 | 29.5 | 33.2 | 29.8 | 28.8 | 30.2 | 33.6 | 35.2 | 34.4 |
| Manufactures | 29.9 | 36.0 | 42.2 | 36.2 | 34.6 | 34.9 | 37.4 | 37.6 | 35.6 |
| Primary products | 24.0 | 24.0 | 20.9 | 22.3 | 22.1 | 28.1 | 33.6 | 35.6 | 36.1 |
| Agricultural products | 20.3 | 21.3 | 17.8 | 19.8 | 18.0 | 23.0 | 27.6 | 28.4 | 28.1 |
| Minerals | 33.5 | 30.7 | 28.6 | 28.5 | 31.4 | 39.0 | 47.3 | 52.7 | 55.3 |
| Services & private transfers | 15.1 | 16.1 | 16.6 | 16.0 | 16.0 | 18.3 | 24.4 | 34.5 | 40.1 |
| Diversified export base | 13.7 | 18.3 | 17.8 | 15.0 | 14.2 | 16.0 | 19.8 | 22.7 | 23.4 |

| By financial criteria | | | | | | | | | |
|---|---|---|---|---|---|---|---|---|---|
| Net creditor countries | 53.7 | 76.4 | 94.4 | 76.1 | 67.2 | 58.3 | 52.0 | 48.8 | 44.3 |
| Net debtor countries | 21.5 | 19.9 | 21.5 | 18.9 | 19.7 | 23.3 | 27.6 | 29.6 | 29.7 |
| Market borrowers | 26.4 | 23.4 | 25.0 | 21.4 | 21.7 | 26.5 | 29.8 | 31.2 | 30.6 |
| Diversified borrowers | 17.1 | 16.3 | 17.3 | 15.4 | 17.3 | 17.5 | 23.5 | 25.5 | 26.0 |
| Official borrowers | 12.5 | 13.5 | 15.6 | 14.5 | 15.4 | 19.0 | 23.4 | 27.7 | 31.1 |
| Countries with recent debt-servicing difficulties | 20.1 | 15.8 | 18.2 | 13.5 | 15.0 | 20.8 | 27.6 | 32.5 | 33.0 |
| Countries without debt-servicing | 22.7 | 23.0 | 23.8 | 22.3 | 22.5 | 24.8 | 27.6 | 28.2 | 28.1 |
| | | | | | | | | | |
| Other groups | | | | | | | | | |
| Small low-income economies | 12.4 | 13.0 | 12.3 | 11.5 | 11.1 | 11.3 | 13.0 | 13.7 | 12.0 |
| Least developed countries | 15.1 | 18.9 | 23.1 | 22.5 | 22.5 | 25.0 | 30.1 | 34.1 | 38.9 |
| 15 heavily indebted countries | 26.8 | 22.6 | 24.1 | 18.1 | 19.4 | 25.9 | 32.4 | 36.5 | 36.3 |

*Source:* IMF, *World Economic Outlook*, Washington, DC, May 1993.
*Notes:* [a] In this table, official holdings of gold are valued at SDR 35 an ounce. This convention results in a marked underestimate of reserves for countries that have gold holdings.
[b] Reserves at year-end in % of imports of goods and services for the year indicated.

sequently. This conclusion appears to hold even when different measures of instability are used, and to apply to Asian and Western Hemisphere developing countries.[6]

## Reserve holdings and adequacy

On the basis of the ratio between international reserves and imports, it transpires that developing countries have generally held more reserves than industrial countries. For the period 1980–9 the average R/M ratio (based on non-gold reserves) for industrial countries was 9.5, whereas that for developing countries was 17.2. However, as Table 1.5 shows, this average conceals significant dispersion amongst developing countries. Yet again it is the low-income countries that emerge as being in the weakest position. The lower holdings of reserves in many of the poorest countries cannot be explained in terms of a lower need for reserves. Since they are faced with payments instability, high adjustment costs, impaired access to international credit and relatively inflexible exchange rates, it can be argued that they need relatively high reserve ratios. While the high opportunity cost of holding reserves encourages such countries to economise on them, there appears to be some *prima facie* evidence of reserve inadequacy.[7] While systemic reserve adequacy is not the problem that it was in the 1960s, reserve adequacy remains an important problem for many developing countries.

This is confirmed by studies that attempt to estimate the future financing needs of developing countries. Clearly such studies need to make assumptions about export and import growth and therefore also implicitly or explicitly about the nature of underlying trade functions. However, the existence of a financing gap for the developing world appears to be quite resilient to modified assumptions concerning both the nature of these functions and future economic growth in the OECD, interest rates and terms of trade. Moreover, given reasonable assumptions about the future availability of commercial credit to many parts of the developing world, it seems likely that the need for finance will translate into a need for official liquidity. Failure to meet this need will force developing countries to earn reserves through payments adjustment.

## Access to international financing

Faced with a balance of payments deficit, countries have two financing options. One is to run down reserves; the other is to borrow from international sources. Tables 1.2 and 1.6 provide information about how developing countries have used these options in the period 1985–93. For developing countries as a whole the period has seen a significant *acquisition* of reserves. Net borrowing from commercial banks was extremely modest during 1985–9 but increased sharply in 1990 and 1991. Borrowing from official creditors has fluctuated less throughout the period and direct investment has shown a persistent increase. However, Table 1.6 shows a stark variation across sub-groups of developing countries as distinguished by regional, financial and other criteria. In particular, it shows how the poorest countries have again been bypassed by the increase in commercial lending at the beginning of the 1990s, as they were in the 1970s.

The swings in the amounts of international financing from different sources have in turn meant big changes in their relative importance. During the 1980s, for example, aid almost doubled in relative importance as many other flows diminished, but it has subsequently become less significant. Moreover, borrowing and financing need to be assessed in relation to the size of balance of payments current account deficits. Whereas net external borrowing of $65.8 bn by developing countries in 1990 more than covered the deficit of $33.4 bn, higher borrowing of $69.3 bn in 1992 coincided with a dramatic increase in the deficit to $96.2 bn. It seems likely that many developing countries will be faced with a significant financing gap throughout the 1990s. As things stand, it seems more likely that this *ex ante* gap will be closed *ex post* by a reduction in the demand for finance associated with the pursuit of contractionary demand management policies in developing countries than by further expansion in foreign aid and direct investment or a sustained reversal in the pattern of bank lending.

It also needs to be recognised that the availability of financing may be adversely affected by capital flight. Although studies have shown the difficulty of putting precise figures on its size, most agree that capital flight has constituted an important element in the financing problems of certain developing countries and that the repatriation of such capital could be of quantitative significance in alleviating the future financing constraints that they face. Attention

Table 1.6 Developing countries: external financing, 1985–93 (US$ bn)

| By region | 1985 | 1986 | 1987 | 1988 | 1989 | 1990 | 1991 | 1992 | 1993 |
|---|---|---|---|---|---|---|---|---|---|
| *Africa* | | | | | | | | | |
| Balance on current account, excluding official transfers | −5.1 | −14.9 | −10.4 | −16.6 | −14.4 | −10.3 | −12.1 | −16.0 | −15.3 |
| Change in reserves (−= increase) | −1.6 | 1.8 | −1.3 | 0.2 | −2.8 | −5.1 | −4.6 | −1.9 | −2.7 |
| *Total net external financing* | 7.9 | 14.6 | 13.0 | 17.3 | 17.2 | 17.0 | 17.1 | 17.2 | 18.4 |
| Non-debt-creating flows, net | 5.0 | 6.8 | 6.8 | 8.3 | 11.2 | 9.4 | 9.6 | 10.0 | 10.5 |
| Reserve-related liabilities | −0.2 | −0.9 | −0.9 | 0.2 | 0.1 | −0.4 | 0.2 | −0.2 | −0.4 |
| Net credit from IMF | 0.1 | −1.0 | −1.1 | −0.3 | 0.1 | −0.6 | 0.2 | −0.2 | – |
| Net external borrowing | 3.1 | 8.7 | 7.2 | 8.8 | 5.8 | 8.0 | 7.3 | 7.4 | 8.3 |
| Net long-term borrowing from official creditors | 4.1 | 6.4 | 6.5 | 6.1 | 5.2 | 9.2 | 6.9 | 8.9 | 11.1 |
| Net borrowing from commercial banks | −0.4 | −1.6 | −1.1 | −0.9 | −2.5 | 2.0 | −0.1 | −4.2 | −2.6 |
| *Asia* | | | | | | | | | |
| Balance on current account, excluding official transfers | −18.7 | −1.1 | 14.8 | 2.6 | −8.1 | −10.0 | −10.2 | −25.2 | −31.2 |
| Change in reserves (−= increase) | −4.0 | −24.4 | −39.5 | −11.3 | −9.0 | −21.6 | −39.3 | −21.9 | −10.5 |
| *Total net external financing* | 24.3 | 29.5 | 27.4 | 19.6 | 21.8 | 36.8 | 58.1 | 50.6 | 51.8 |
| Non-debt-creating flows, net | 6.6 | 8.5 | 10.5 | 11.4 | 9.1 | 12.8 | 16.8 | 20.1 | 21.9 |
| Reserve-related liabilities | −0.9 | −0.9 | −2.4 | −2.5 | −1.2 | −2.3 | 2.8 | 0.5 | 0.6 |
| Net credit from IMF | −1.0 | −0.9 | −2.4 | −2.4 | −1.1 | −2.4 | 1.9 | 1.3 | – |
| Net external borrowing | 18.6 | 21.9 | 19.2 | 10.7 | 13.9 | 26.3 | 38.5 | 29.9 | 29.3 |
| Net long-term borrowing from official creditors | 7.5 | 8.4 | 7.5 | 7.1 | 8.3 | 13.1 | 10.8 | 10.6 | 8.7 |
| Net borrowing from commercial banks | 8.8 | 4.7 | 5.0 | 4.2 | −0.7 | 7.7 | 20.4 | 7.2 | 8.3 |

*Middle East & Europe*

| | | | | | | | | | |
|---|---|---|---|---|---|---|---|---|---|
| Balance on current account, excluding official transfers | −13.2 | −28.7 | −17.1 | −16.4 | −5.5 | −4.6 | −35.7 | −20.3 | −2.9 |
| Change in reserves (− = increase) | −4.4 | 5.9 | −6.5 | 6.8 | −10.9 | −8.1 | −4.6 | −7.3 | −8.8 |
| *Total net external financing* | *7.5* | *18.8* | *13.3* | *7.0* | *10.2* | *9.1* | *−0.2* | *22.0* | *9.1* |
| Non-debt-creating flows, net | 5.5 | 4.2 | 6.7 | 2.8 | 4.1 | 0.3 | −13.5 | 8.2 | 0.8 |
| Reserve-related liabilities | −0.2 | −0.5 | −0.4 | −0.5 | −0.2 | −0.1 | − | 0.4 | − |
| Net credit from IMF | −0.2 | −0.5 | −0.4 | −0.5 | −0.2 | −0.1 | − | 0.4 | − |
| Net external borrowing | 2.3 | 15.1 | 6.9 | 4.8 | 6.3 | 8.9 | 13.3 | 13.4 | 8.3 |
| Net long-term borrowing from official creditors | 4.4 | 5.3 | 7.1 | 2.4 | 4.3 | 5.3 | 5.5 | 14.1 | 6.5 |
| Net borrowing from commercial banks | 2.2 | 2.5 | 6.1 | 6.2 | −3.7 | 3.2 | 3.6 | −1.1 | 0.4 |

*Western Hemisphere*

| | | | | | | | | | |
|---|---|---|---|---|---|---|---|---|---|
| Balance on current account, excluding official transfers | −5.5 | −19.8 | −11.8 | −13.4 | −10.1 | −8.5 | −20.5 | −34.7 | −35.1 |
| Change in reserves (− = increase) | −1.0 | 6.6 | −3.2 | 7.7 | −2.6 | −15.1 | −18.4 | −18.7 | −4.0 |
| *Total net external financing* | *13.2* | *16.1* | *14.4* | *12.4* | *19.8* | *30.2* | *30.6* | *33.6* | *26.8* |
| Non-debt-creating flows, net | 5.6 | 4.6 | 4.7 | 8.7 | 7.9 | 8.9 | 13.8 | 15.8 | 15.2 |
| Reserve-related liabilities | 2.0 | 1.5 | − | −0.4 | 0.2 | −1.2 | −1.4 | −0.8 | −0.9 |
| Net credit from IMF | 1.5 | 0.1 | −0.8 | −0.9 | −0.2 | 1.2 | −1.0 | −1.6 | − |
| Net external borrowing | 5.5 | 10.0 | 9.7 | 4.1 | 11.7 | 22.6 | 18.2 | 18.6 | 12.4 |
| Net long-term borrowing from official creditors | 8.1 | 10.8 | 5.8 | 6.1 | 10.8 | 20.0 | 7.4 | 8.7 | 9.5 |
| Net borrowing from commercial banks | −5.8 | −4.2 | −1.8 | −7.0 | 5.2 | 13.5 | 12.2 | 3.5 | 10.1 |

Table 1.6 (continued)

|  | 1985 | 1986 | 1987 | 1988 | 1989 | 1990 | 1991 | 1992 | 1993 |
|---|---|---|---|---|---|---|---|---|---|
| *Sub-Saharan Africa* | | | | | | | | | |
| Balance on current account, excluding official transfers | −7.1 | −10.0 | −11.1 | −12.9 | −12.4 | −14.3 | −15.1 | −16.4 | −16.0 |
| Change in reserves (−= increase) | −0.6 | −0.6 | −0.2 | −1.1 | −1.0 | −0.9 | −1.7 | −2.1 | −1.9 |
| *Total net external financing* | *8.5* | *11.0* | *12.4* | *14.3* | *13.3* | *15.2* | *16.3* | *18.0* | *18.0* |
| Non-debt-creating flows, net | 4.2 | 5.2 | 5.4 | 6.2 | 6.7 | 6.7 | 6.9 | 7.5 | 7.8 |
| Reserve-related liabilities | 0.1 | −0.5 | −0.2 | 0.3 | −0.4 | −0.2 | − | 0.1 | 0.1 |
| Net credit from IMF | − | −0.4 | −0.5 | −0.2 | −0.4 | −0.3 | − | − | − |
| Net external borrowing | 4.2 | 6.3 | 7.2 | 7.8 | 7.0 | 8.7 | 9.5 | 10.4 | 10.1 |
| Net long-term borrowing from official creditors | 3.8 | 5.7 | 6.6 | 5.6 | 3.8 | 9.0 | 6.7 | 6.4 | 8.3 |
| Net borrowing from commercial banks | 0.5 | −0.5 | −0.1 | − | 0.3 | 1.3 | 0.6 | 0.6 | 0.6 |
| *Four newly industrialising Asian economies* | | | | | | | | | |
| Balance on current account, excluding official transfers | 8.3 | 21.2 | 27.7 | 25.2 | 18.9 | 10.8 | 7.8 | −1.5 | −2.9 |
| Change in reserves (−= increase) | −8.5 | −25.1 | −33.0 | −10.3 | −5.6 | −3.3 | −13.3 | −8.2 | 2.6 |
| *Total net external financing* | *2.1* | *6.4* | *6.0* | *−8.3* | *−8.7* | *−7.3* | *12.1* | *8.4* | *2.2* |
| Non-debt-creating flows, net | 1.3 | 2.1 | 3.0 | 1.1 | −2.6 | −0.7 | 1.8 | 1.8 | 1.6 |
| Reserve-related liabilities | −0.2 | −0.1 | −1.2 | −0.5 | − | − | − | − | − |
| Net credit from IMF | −0.2 | −0.1 | −1.2 | −0.5 | − | − | − | − | − |
| Net external borrowing | 1.1 | 4.4 | 4.1 | −8.9 | −6.1 | −0.6 | 10.3 | 6.6 | 0.6 |
| Net long-term borrowing from official creditors | − | −0.2 | −2.4 | −0.9 | −0.3 | 0.2 | 0.9 | 0.4 | 0.3 |
| Net borrowing from commercial banks | 2.9 | 2.4 | −0.4 | −4.9 | −5.4 | −6.5 | 8.2 | 2.4 | 1.7 |

*By predominant export*

**Fuel**

| | | | | | | | | | |
|---|---:|---:|---:|---:|---:|---:|---:|---:|---:|
| Balance on current account, excluding official transfers | 0.3 | −36.8 | −11.1 | −27.8 | −9.5 | 5.2 | −42.4 | −43.9 | −26.1 |
| Change in reserves (−= increase) | −3.5 | 15.6 | −9.7 | 20.1 | −8.7 | −15.4 | −16.3 | −1.6 | −0.6 |
| *Total net external financing* | −0.5 | 22.2 | 11.6 | 7.6 | 11.3 | 9.9 | 10.0 | 25.2 | 17.5 |
| Non-debt-creating flows, net | −0.7 | — | 2.3 | 0.9 | 4.3 | −4.5 | −14.6 | 9.9 | 4.5 |
| Reserve-related liabilities | — | 1.7 | 0.3 | 0.9 | 3.2 | 1.4 | 0.3 | −1.3 | −2.0 |
| Net credit from IMF | — | 0.8 | 1.0 | — | 2.0 | 2.7 | 0.3 | −1.3 | — |
| Net external borrowing | 0.2 | 20.5 | 9.0 | 5.8 | 3.7 | 12.9 | 24.3 | 16.7 | 15.0 |
| Net long-term borrowing from official creditors | 9.0 | 8.1 | 12.5 | 8.5 | 5.8 | 14.4 | 9.0 | 14.2 | 10.3 |
| Net borrowing from commercial banks | −0.7 | −1.1 | 0.9 | −2.6 | 0.4 | 17.3 | 16.8 | 3.4 | 7.2 |

**Non-fuel exports**

| | | | | | | | | | |
|---|---:|---:|---:|---:|---:|---:|---:|---:|---:|
| Balance on current account, excluding official transfers | −42.7 | −27.7 | −13.4 | −15.9 | −28.6 | −38.5 | −36.1 | −52.2 | −58.4 |
| Change in reserves (−= increase) | −7.6 | −25.6 | −40.7 | −16.6 | −16.6 | −34.5 | −50.5 | −48.2 | −25.3 |
| *Total net external financing* | 53.4 | 56.8 | 56.5 | 48.8 | 57.7 | 83.3 | 95.6 | 98.0 | 88.4 |
| Non-debt-creating flows, net | 23.5 | 24.1 | 26.5 | 30.3 | 28.0 | 35.9 | 41.3 | 44.2 | 43.8 |
| Reserve-related liabilities | 0.7 | −2.5 | −4.0 | −4.1 | −4.2 | −5.5 | 1.3 | 1.3 | 1.3 |
| Net credit from IMF | 0.3 | −3.0 | −5.7 | −4.1 | −3.5 | −4.6 | 1.2 | 1.2 | 1.3 |
| Net external borrowing | 29.2 | 35.3 | 34.0 | 22.6 | 33.9 | 52.9 | 53.1 | 52.6 | 43.3 |
| Net long-term borrowing from official creditors | 15.1 | 23.0 | 14.4 | 13.2 | 22.8 | 33.2 | 21.7 | 28.1 | 25.4 |
| Net borrowing from commercial banks | 5.6 | 2.5 | 7.3 | 5.1 | −2.1 | 9.1 | 19.3 | 2.1 | 9.0 |

Table 1.6 (continued)

| By financial criteria | 1985 | 1986 | 1987 | 1988 | 1989 | 1990 | 1991 | 1992 | 1993 |
|---|---|---|---|---|---|---|---|---|---|
| *Net creditor countries* | | | | | | | | | |
| Balance on current account, excluding official transfers | 14.5 | 8.9 | 15.4 | 7.8 | 16.3 | 23.7 | −13.6 | −13.7 | 0.1 |
| Change in reserves (−= increase) | −11.2 | −19.5 | −34.8 | 7.6 | −2.1 | 2.1 | −11.9 | −0.9 | 3.9 |
| *Total net external financing* | *−6.6* | *14.3* | *15.7* | *−10.6* | *−10.2* | *−20.5* | *−14.9* | *7.1* | *−7.4* |
| Non-debt-creating flows, net | −2.2 | −3.3 | −0.5 | −6.5 | −8.6 | −14.3 | −25.3 | −1.2 | −6.9 |
| Reserve-related liabilities | – | – | – | – | – | – | – | – | – |
| Net credit from IMF | – | – | – | – | – | – | – | – | – |
| Net external borrowing | −4.4 | 17.6 | 16.1 | −4.1 | −1.5 | −6.2 | 10.4 | 8.3 | −0.5 |
| Net long-term borrowing from official creditors | −0.9 | −0.8 | −0.7 | 0.1 | 0.1 | 0.6 | 3.1 | 9.5 | 3.1 |
| Net borrowing from commercial banks | 0.6 | 2.7 | 9.4 | 0.5 | −4.5 | −7.6 | 5.7 | −0.9 | −1.1 |
| *Net debtor countries* | | | | | | | | | |
| Balance on current account, excluding official transfers | −57.0 | −73.4 | −40.0 | −51.6 | −54.4 | −57.1 | −65.0 | −82.5 | −84.6 |
| Change in reserves (−= increase) | 0.1 | 9.5 | −15.6 | −4.1 | −23.1 | −52.0 | −55.0 | −49.0 | −29.8 |
| *Total net external financing* | *59.5* | *64.8* | *52.4* | *66.9* | *79.2* | *113.6* | *120.6* | *116.2* | *113.4* |
| Non-debt-creating flows, net | 25.0 | 27.3 | 29.2 | 37.7 | 41.0 | 45.7 | 52.1 | 55.2 | 55.2 |
| Reserve-related liabilities | 0.7 | −0.8 | −3.7 | −3.2 | −1.0 | −4.0 | 1.6 | – | −0.7 |
| Net credit from IMF | 0.3 | −2.2 | −4.7 | −4.1 | −1.5 | −1.9 | 1.1 | −0.2 | – |
| Net external borrowing | 33.8 | 38.2 | 26.9 | 32.4 | 39.1 | 72.0 | 66.9 | 61.0 | 58.9 |
| Net long-term borrowing from official creditors | 25.0 | 31.9 | 27.6 | 21.5 | 28.5 | 47.0 | 27.5 | 32.8 | 32.7 |
| Net borrowing from commercial banks | 4.3 | −1.3 | −1.2 | 2.1 | 2.8 | 34.0 | 30.4 | 6.4 | 17.3 |

*Market borrowers*

| | | | | | | | | | |
|---|---|---|---|---|---|---|---|---|---|
| Balance on current account, excluding official transfers | −19.1 | −26.3 | −0.8 | −5.7 | −10.9 | −6.0 | −27.4 | −41.6 | −44.6 |
| Change in reserves (−= increase) | 0.2 | 7.1 | −12.8 | −3.8 | −14.0 | −36.1 | −37.7 | −34.1 | −16.2 |
| *Total net external financing* | *27.6* | *24.5* | *12.5* | *22.9* | *34.8* | *57.9* | *71.1* | *60.9* | *58.0* |
| Non-debt-creating flows, net | 11.6 | 13.0 | 14.1 | 19.8 | 20.0 | 22.7 | 28.2 | 31.7 | 32.5 |
| Reserve-related liabilities | 1.7 | 2.0 | −1.1 | −1.1 | 0.6 | −1.7 | −1.4 | −0.8 | −1.0 |
| Net credit from IMF | 1.2 | 0.6 | −1.8 | −1.4 | 0.2 | 0.7 | −1.2 | −1.6 | – |
| Net external borrowing | 14.3 | 9.5 | −0.5 | 4.2 | 14.2 | 36.8 | 44.3 | 30.0 | 26.5 |
| Net long-term borrowing from official creditors | 6.9 | 9.0 | −0.1 | 4.1 | 8.6 | 20.0 | 8.8 | 9.5 | 11.1 |
| Net borrowing from commercial banks | 0.3 | −6.3 | −5.4 | −3.1 | 7.9 | 22.8 | 27.6 | 9.8 | 15.2 |

*Diversified borrowers*

| | | | | | | | | | |
|---|---|---|---|---|---|---|---|---|---|
| Balance on current account, excluding official transfers | −19.2 | −22.6 | −18.2 | −21.9 | −20.1 | −28.5 | −14.6 | −19.0 | −17.5 |
| Change in reserves (−= increase) | – | 3.7 | −1.0 | 2.0 | −6.4 | −4.7 | −11.2 | −7.9 | −6.3 |
| *Total net external financing* | *14.0* | *16.0* | *16.4* | *17.8* | *18.3* | *23.9* | *22.0* | *26.6* | *25.6* |
| Non-debt-creating flows, net | 5.0 | 4.9 | 4.3 | 6.5 | 6.9 | 7.9 | 9.5 | 10.1 | 9.0 |
| Reserve-related liabilities | −0.3 | −2.1 | −1.8 | −1.3 | −1.7 | −1.2 | 2.6 | 0.3 | 0.6 |
| Net credit from IMF | −0.7 | −1.7 | −2.2 | −1.9 | −1.5 | −1.5 | 2.0 | 1.1 | – |
| Net external borrowing | 9.3 | 13.2 | 13.9 | 12.6 | 13.2 | 17.3 | 9.8 | 16.3 | 16.0 |
| Net long-term borrowing from official creditors | 9.0 | 10.1 | 16.4 | 4.9 | 6.5 | 13.1 | 8.2 | 8.6 | 8.2 |
| Net borrowing from commercial banks | 4.8 | 6.0 | 5.0 | 5.5 | −3.4 | 11.1 | 2.8 | 0.1 | 1.9 |

Table 1.6 (continued)

| | 1985 | 1986 | 1987 | 1988 | 1989 | 1990 | 1991 | 1992 | 1993 |
|---|---|---|---|---|---|---|---|---|---|
| *Official borrowers* | | | | | | | | | |
| Balance on current account, excluding official transfers | −18.6 | −24.5 | −21.0 | −24.0 | −23.4 | −22.6 | −23.0 | −21.9 | −22.5 |
| Change in reserves (−= increase) | −0.1 | −1.4 | −1.8 | −2.4 | −2.7 | −11.2 | −6.1 | −7.0 | −7.4 |
| *Total net external financing* | 18.0 | 24.3 | 23.5 | 26.2 | 26.0 | 31.8 | 27.6 | 28.6 | 29.9 |
| Non-debt-creating flows, net | 8.4 | 9.4 | 10.8 | 11.4 | 14.1 | 15.1 | 14.3 | 13.5 | 13.7 |
| Reserve-related liabilities | −0.6 | −0.7 | −0.8 | −0.8 | 0.1 | −1.1 | 0.4 | 0.4 | −0.3 |
| Net credit from IMF | −0.2 | −1.0 | −0.8 | −0.8 | −0.2 | −1.1 | 0.3 | 0.3 | – |
| Net external borrowing | 10.2 | 15.5 | 13.5 | 15.6 | 11.8 | 17.9 | 12.9 | 14.7 | 16.4 |
| Net long-term borrowing from official creditors | 9.1 | 12.7 | 11.3 | 12.5 | 13.4 | 13.9 | 10.5 | 14.7 | 13.4 |
| Net borrowing from commercial banks | −0.8 | −1.0 | −0.8 | −0.4 | −1.7 | 0.2 | −0.1 | −3.5 | 0.2 |
| *Countries with recent debt-servicing difficulties* | | | | | | | | | |
| Balance on current account, excluding official transfers | −22.5 | −48.2 | −31.3 | −38.4 | −31.1 | −31.0 | −38.3 | −52.2 | −50.9 |
| Change in reserves (−= increase) | −0.1 | 9.9 | −6.8 | 7.0 | −10.0 | −28.1 | −24.3 | −22.4 | −9.9 |
| *Total net external financing* | 22.4 | 33.0 | 32.5 | 32.2 | 39.7 | 55.6 | 53.0 | 54.9 | 49.0 |
| Non-debt-creating flows, net | 11.1 | 11.9 | 13.4 | 17.9 | 18.9 | 20.2 | 24.5 | 26.1 | 24.9 |
| Reserve-related liabilities | 1.8 | 0.5 | −0.7 | −0.4 | – | −1.8 | −0.5 | −1.7 | −1.2 |
| Net credit from IMF | 1.7 | −1.0 | −1.8 | −1.3 | −0.5 | 0.4 | −1.0 | −1.8 | – |
| Net external borrowing | 9.5 | 20.6 | 19.9 | 14.7 | 20.8 | 37.2 | 29.0 | 30.5 | 25.3 |
| Net long-term borrowing from official creditors | 15.8 | 19.0 | 18.3 | 15.6 | 21.5 | 34.1 | 15.9 | 22.2 | 20.6 |
| Net borrowing from commercial banks | −5.6 | −5.3 | −1.0 | −4.0 | – | 15.0 | 9.6 | −4.3 | 9.3 |

*Countries without debt-servicing difficulties*

| | | | | | | | | | |
|---|---|---|---|---|---|---|---|---|---|
| Balance on current account, excluding official transfers | −34.4 | −25.2 | −8.7 | −13.2 | −23.3 | −26.1 | −26.7 | −30.3 | −33.7 |
| Change in reserves (−= increase) | 0.2 | −0.5 | −8.8 | −11.1 | −13.2 | −24.0 | −30.7 | −26.6 | −19.9 |
| *Total net external financing* | *37.1* | *31.8* | *19.9* | *34.7* | *39.4* | *58.0* | *67.6* | *61.2* | *64.4* |
| Non-debt-creating flows, net | 13.8 | 15.5 | 15.9 | 19.8 | 22.1 | 25.5 | 27.6 | 29.1 | 30.3 |
| Reserve-related liabilities | −1.0 | −1.3 | −3.0 | −2.8 | −1.0 | −2.3 | 2.1 | 1.6 | 0.5 |
| Net credit from IMF | −1.4 | −1.2 | −2.9 | −2.8 | −1.0 | −2.3 | 2.1 | 1.6 | – |
| Net external borrowing | 24.3 | 17.6 | 7.0 | 17.7 | 18.3 | 34.7 | 37.9 | 30.5 | 33.6 |
| Net long-term borrowing from official creditors | 9.2 | 12.9 | 9.3 | 5.9 | 7.0 | 12.9 | 11.6 | 10.7 | 12.0 |
| Net borrowing from commercial banks | 9.8 | 4.1 | −0.2 | 6.1 | 2.8 | 19.1 | 20.8 | 10.7 | 8.0 |

*Small low-income countries*

| | | | | | | | | | |
|---|---|---|---|---|---|---|---|---|---|
| Balance on current account, excluding official transfers | −12.3 | −13.0 | −14.4 | −16.4 | −17.3 | −18.4 | −18.8 | −19.4 | −19.5 |
| Change in reserves (−= increase) | 0.7 | −0.3 | −0.1 | −0.3 | −0.5 | – | −0.2 | −0.4 | −0.3 |
| *Total net external financing* | *11.9* | *13.7* | *14.9* | *17.1* | *17.9* | *18.2* | *19.2* | *20.1* | *20.0* |
| Non-debt-creating flows, net | 5.3 | 6.2 | 6.8 | 7.4 | 7.8 | 8.1 | 8.5 | 8.7 | 8.9 |
| Reserve-related liabilities | 0.1 | −0.8 | −0.4 | −0.3 | – | −0.5 | 0.5 | 0.3 | 0.2 |
| Net credit from IMF | −0.2 | −0.9 | −0.6 | −0.3 | – | −0.6 | 0.4 | 0.2 | – |
| Net external borrowing | 6.5 | 8.3 | 8.6 | 10.0 | 10.1 | 10.7 | 10.2 | 11.1 | 11.1 |
| Net long-term borrowing from official creditors | 3.8 | 5.9 | 5.3 | 5.7 | 7.5 | 9.3 | 7.4 | 9.0 | 8.2 |
| Net borrowing from commercial banks | 0.6 | −0.4 | 0.5 | 0.4 | −0.1 | 1.1 | – | 0.1 | 0.4 |

*Table 1.6* (continued)

| | 1985 | 1986 | 1987 | 1988 | 1989 | 1990 | 1991 | 1992 | 1993 |
|---|---|---|---|---|---|---|---|---|---|
| *Least developed countries* | | | | | | | | | |
| Balance on current account, excluding official transfers | -9.1 | -9.7 | -10.3 | -11.4 | -11.3 | -13.0 | -13.1 | -13.6 | -13.6 |
| Change in reserves (-= increase) | -0.1 | -1.1 | -1.3 | -1.1 | -0.6 | -0.8 | -1.8 | -2.5 | -2.7 |
| *Total net external financing* | *9.5* | *11.3* | *12.2* | *13.2* | *12.3* | *13.9* | *15.0* | *16.3* | *16.4* |
| Non-debt-creating flows, net | 4.6 | 5.6 | 5.8 | 6.2 | 6.5 | 7.1 | 7.3 | 7.7 | 7.9 |
| Reserve-related liabilities | – | -0.2 | 0.2 | -0.1 | -0.4 | -0.3 | 0.1 | 0.3 | 0.2 |
| Net credit from IMF | -0.1 | -0.3 | – | -0.2 | -0.3 | -0.4 | 0.1 | 0.2 | – |
| Net external borrowing | 4.9 | 5.9 | 6.2 | 7.1 | 6.2 | 7.1 | 7.6 | 8.3 | 8.3 |
| Net long-term borrowing from official creditors | 3.4 | 4.9 | 4.8 | 4.7 | 4.2 | 5.9 | 4.6 | 5.8 | 5.9 |
| Net borrowing from commercial banks | 0.5 | -0.3 | 0.3 | -0.4 | 0.5 | 0.7 | 0.1 | – | 0.3 |
| *15 heavily indebted countries* | | | | | | | | | |
| Balance on current account, excluding official transfers | -2.0 | -19.3 | -9.3 | -11.6 | -7.5 | -5.8 | -23.2 | -34.0 | -35.0 |
| Change in reserves (-= increase) | -1.7 | 5.7 | -2.5 | 5.8 | -7.0 | -18.6 | -17.5 | -16.3 | -5.5 |
| *Total net external financing* | *10.4* | *16.5* | *12.4* | *11.6* | *16.8* | *31.6* | *32.0* | *31.2* | *28.9* |
| Non-debt-creating flows, net | 5.1 | 4.2 | 4.1 | 8.8 | 10.0 | 9.6 | 14.0 | 16.1 | 16.0 |
| Reserve-related liabilities | 1.7 | 1.1 | -0.4 | -0.7 | -0.3 | -1.7 | -0.7 | -1.8 | -1.1 |
| Net credit from IMF | 1.6 | -0.2 | -1.3 | -1.4 | -0.8 | 0.6 | -1.4 | -1.8 | – |
| Net external borrowing | 3.6 | 11.2 | 8.6 | 3.5 | 7.1 | 23.7 | 18.7 | 16.8 | 14.0 |
| Net long-term borrowing from official creditors | 11.4 | 13.0 | 7.4 | 6.8 | 13.1 | 22.0 | 10.0 | 10.9 | 10.4 |
| Net borrowing from commercial banks | -7.7 | -4.8 | -3.5 | -6.4 | 3.4 | 13.9 | 11.0 | -2.9 | 9.1 |

*Source:* IMF, *World Economic Outlook*, Washington, DC, May 1993.

may therefore need to be paid to the design of preferential exchange rate and taxation schemes and measures to raise relative creditworthiness and the risk-adjusted expected net yield on domestic as compared with foreign assets.

## Debt

The debt situation for much of the developing world is summarised in Table 1.7. A conventional debt indicator such as the debt-service ratio provides some evidence that the problem has been ameliorated for Latin American debtors. But for African debtors it has shown no inclination to fall since 1987, and by 1993 was higher than for developing countries in the Western Hemisphere.

Moreover, the debt situation has affected economic performance in indebted countries. The pursuit of adjustment programmes has often meant a sharp decline in the rate of economic growth and, given population growth, falling or stagnating living standards. Investment has not been insulated from adjustment, with domestic saving being channelled into servicing external debt. Future growth performance has therefore been undermined. A strengthening in the balance of payments has often been engineered by inducing a cut-back in imports that exceeds the fall in exports, even though imports constitute vital inputs into the generation of economic growth and exports, and therefore an economy's future capacity to service debt.

While even with a relatively low level of borrowing from private capital markets during the 1970s and 1980s low-income countries have not escaped debt problems, it must also be recognised that the debt indicators taken in isolation may be misleading (see, for example, Bird, 1985). Thus, although in terms of debt ratios, low-income countries have often encountered more acute problems than those of the highly indebted countries, such comparisons may overstate the relative size of their problems given that a high proportion of the debt is official. Nevertheless their balance of payments and economic development have been severely constrained.

## The scope for and cost of adjustment

The factors discussed above combine to suggest that, compared with developed countries, many developing countries will be more

Table 1.7 Developing countries: debt and debt-service ratios (in % of exports of goods and services)

| External debt of developing countries | 1985 | 1986 | 1987 | 1988 | 1989 | 1990 | 1991 | 1992 | 1993 |
|---|---|---|---|---|---|---|---|---|---|
| | 154.7 | 180.1 | 166.2 | 147.4 | 134.6 | 126.9 | 126.8 | 119.9 | 112.3 |
| *By region* | | | | | | | | | |
| Africa | 186.9 | 241.4 | 245.0 | 244.8 | 237.1 | 219.6 | 230.4 | 229.1 | 226.7 |
| Asia | 104.9 | 107.5 | 92.3 | 78.1 | 71.0 | 69.8 | 69.3 | 64.7 | 61.4 |
| Middle East & Europe | 107.2 | 153.8 | 150.3 | 151.1 | 136.4 | 124.4 | 134.6 | 132.8 | 125.6 |
| Western Hemisphere | 292.5 | 346.3 | 336.9 | 292.8 | 268.4 | 250.9 | 257.9 | 249.7 | 231.4 |
| *By financial criteria* | | | | | | | | | |
| Net creditor countries | 35.4 | 44.9 | 42.0 | 39.8 | 32.5 | 26.7 | 31.6 | 32.3 | 30.2 |
| Net debtor countries | 192.9 | 219.3 | 201.3 | 175.5 | 162.4 | 154.5 | 152.0 | 141.8 | 131.9 |
| Market borrowers | 168.8 | 179.0 | 154.1 | 124.6 | 111.3 | 104.3 | 102.4 | 95.2 | 88.2 |
| Diversified borrowers | 193.6 | 236.6 | 241.7 | 232.7 | 223.1 | 227.8 | 234.0 | 223.3 | 204.1 |
| Official borrowers | 305.9 | 388.8 | 384.3 | 372.9 | 358.2 | 319.1 | 321.7 | 313.2 | 304.6 |
| Countries with recent debt-servicing difficulties | 278.1 | 347.9 | 336.8 | 311.7 | 290.1 | 276.6 | 288.9 | 280.6 | 258.9 |
| Countries without debt-servicing difficulties | 123.4 | 132.4 | 118.5 | 100.1 | 92.5 | 90.8 | 89.1 | 82.7 | 78.5 |

| Debt service payments of developing countries | 20.9 | 22.8 | 20.6 | 19.2 | 16.2 | 14.2 | 14.4 | 14.8 | 14.2 |
|---|---|---|---|---|---|---|---|---|---|
| *By region* | | | | | | | | | |
| Africa | 27.5 | 28.0 | 23.7 | 24.7 | 24.1 | 23.3 | 25.2 | 26.6 | 33.4 |
| Asia | 14.1 | 15.0 | 14.7 | 11.2 | 10.3 | 8.9 | 7.7 | 7.4 | 7.3 |
| Middle East & Europe | 11.8 | 16.7 | 16.0 | 16.2 | 16.2 | 14.0 | 14.5 | 13.1 | 12.8 |
| Western Hemisphere | 41.8 | 45.4 | 40.5 | 43.1 | 30.0 | 25.8 | 31.7 | 37.8 | 32.2 |
| *By financial criteria* | | | | | | | | | |
| Net creditor countries | 5.3 | 7.4 | 6.5 | 6.5 | 5.8 | 5.2 | 5.8 | 5.2 | 5.2 |
| Net debtor countries | 25.9 | 27.3 | 24.6 | 22.5 | 19.1 | 16.7 | 16.7 | 17.2 | 16.3 |
| Market borrowers | 25.7 | 26.7 | 23.8 | 20.8 | 15.8 | 13.2 | 13.6 | 14.5 | 12.9 |
| Diversified borrowers | 25.4 | 26.2 | 25.3 | 27.5 | 26.3 | 25.6 | 24.1 | 23.9 | 22.2 |
| Official borrowers | 27.8 | 32.4 | 28.0 | 21.9 | 24.4 | 21.0 | 23.3 | 23.2 | 29.2 |
| Countries with recent debt-servicing difficulties | 35.6 | 38.6 | 33.7 | 34.4 | 27.0 | 24.0 | 27.6 | 31.7 | 30.2 |
| Countries without debt-servicing difficulties | 18.0 | 19.6 | 19.1 | 15.9 | 14.7 | 12.9 | 11.8 | 11.1 | 10.5 |

*Source:* IMF, *World Economic Outlook*, Washington, DC, May 1993.

likely to encounter balance of payments deficits, and will find it more difficult to finance these either by running down reserves or by international borrowing. They will therefore be under relatively great pressure to eliminate their deficits through adjustment. But what is the scope for adjustment within developing countries, and at what cost can it be achieved?

There is a reasonable presumption that most poor countries possess a relatively low degree of structural flexibility. Markets may often be ill developed and price elasticities low, with the result that the scope for switching resources rapidly into the production of traded goods will be strictly limited. In a global economic environment hostile to export expansion, developing country governments possess few alternatives to a deflationary programme of balance of payments stabilisation. The costs of such programmes in economic growth and future export performance have already been mentioned. In addition, the imposition of such economic costs has been shown to put considerable strain on fragile democratic political systems, and it is important therefore not to lose sight of the political costs of adjustment.[8] The capacity for short-run adjustment is likely to be particularly constrained in low-income countries where economies are inflexible, *per capita* income is at a low level, technical competence is limited, and political support for the government is tenuous.

An overall assessment of the above criteria again reveals that it is unwise and inappropriate to group all developing countries together. Different countries perform differently under different criteria. However, the evidence does suggest that many parts of the developing world, and perhaps in particular the poorest parts, have experienced: structural weakening in their balance of payments; instability associated with export concentration; low levels of both reserves and access to finance; and severe adjustment problems. There is thus justification for treating such countries as a special case within the international monetary system – an argument that is only enhanced by considerations of international equity.

Although the early 1980s were dominated by concerns about the financing and adjustment problems facing middle-income countries, concern has more recently also been directed towards what are the often more intractable problems facing the poorest of the poor countries. As the empirical evidence quoted in this chapter shows, the IMF has been involved in both groups.

# THE HIGHLY INDEBTED DEVELOPING COUNTRIES AND THE FUND'S INVOLVEMENT

While losing many of its systemic functions, the Fund's operations during the 1980s became dominated by dealing with the debt difficulties faced by a relatively small group of highly indebted developing countries. All the Fund's lending was to developing countries, and the majority of it was to the highly indebted countries, even though the majority of programmes remained with low-income countries.

The Fund frequently became depicted as a development agency offering concessional assistance to developing countries. Even some of its staff bemoaned what they saw as the loss of its monetary characteristics and consequently much of its financial reputation (Finch, 1988). The least subtle criticisms of this type tended to use the phrase 'development agency' almost as a term of abuse. What the Fund was doing was perceived as being bad in and of itself. The more subtle criticism was that the Fund had largely been pushed by political pressure into lowering its own financial standards.

The criticism here was not so much that development assistance is inappropriate, but rather that the IMF is an inappropriate institution through which to give it. This argument sees it as important to retain the revolving character of Fund resources, as well as the Fund's short-term monetary perspective – features, so it is claimed, that will be lost if the Fund is forced to lend over the long term on the basis of unviable programmes and unachievable targets. The plea has been strongly articulated to 'let the IMF be the IMF' (Finch, 1988). An extension of this argument is that unsuccessful programmes will damage the reputation and credibility of the Fund and adversely affect its catalytic role.

The claim that financial standards have been sacrificed is intimately related to the debt crisis. In essence, it is that the governments of countries where the private banks are located, and in particular the United States, encouraged the Fund to lend to the highly indebted countries in order to reduce the probability of default. In the early years of the debt crisis, the argument could be made that such action was sustaining the stability of the international banking system. But as the banks themselves adjusted to the crisis by reducing their exposure, strengthening their capital adequacy, provisioning, and expanding other lines of business, this systemic argument for lending by the IMF disappeared.

Even critics who approach the issue from a rather different angle, having more in common with the 'traditional' criticisms of Fund conditionality, have concluded that the main beneficiaries of Fund lending to highly indebted developing countries during the 1980s were the international banks. Simply put, the claim is that it was positive net transfers from the Fund that financed negative net transfers with the banks. This is a claim that is at least superficially consistent with the evidence at aggregate level, but it is not an interpretation that finds ready acceptance – publicly at least – inside the Fund, where the accusation that it had bailed out the banks has been, often staunchly, rejected.[9]

Yet the criticism that the Fund failed in its dealings with the highly indebted countries during the 1980s has more dimensions to it than this. First, there is the argument that, along with others, the Fund misinterpreted the very nature of the debt crisis by treating it either as a liquidity crisis or as one of short-term internal adjustment rather than as a more deep-seated problem of structural adjustment which required important supply-side responses as well as the appropriate management of demand. This meant that the Fund opted to support new financing which assisted countries in meeting their outstanding debt-servicing obligations but which did little to restore medium-term viability to their balance of payments.

The nature of the programmes supported by the Fund has, in relation to this, been criticised for an overemphasis on devaluation resulting from a desire to strengthen the tradeables sector of the economy and thereby to facilitate debt servicing, and an over-ambitious attempt to achieve stabilisation and liberalisation simultaneously.[10] A long-standing worry associated with the use of devaluation is that a shift in the nominal exchange rate will fail to alter the real exchange rate because of the inflation it generates. Devaluation is seen as destroying the 'nominal anchor', or to use the older jargon 'reserve discipline', that a fixed exchange rate provides. Is this not a particular worry in highly indebted countries where the inflation record is frequently very poor and where the reputation of governments as inflation fighters is often weak? Just as the counter-inflationary merits of fixed exchange rates were being acknowledged and accentuated in the context of the

38

European Monetary System, were they not being neglected by the IMF?

Critics of the Fund's approach to conditionality within the highly indebted countries have argued that, whereas devaluation may certainly be appropriate in some circumstances (where, for example, it is designed to unify the exchange rate, respond to an external shock that has altered the equilibrium real exchange rate, or negate the impact of monetised fiscal deficits on the external account) it may be inappropriate where the fiscal deficit is under control and where the income redistributive effects, particularly in terms of lowering the urban real wage, spark off political unrest and measures to restore real wages. In these circumstances, the price of non-tradeables may also rise, with the result that the relative price effect of devaluation on the internal terms of trade is lost. The dangers of a vicious circle, whereby inflation leads to devaluation which then leads to further inflation, have long been acknowledged in Latin American economies where there is a legacy of rapid inflation and a low degree of money illusion.[11] Indeed, in the context of forward-looking models of economic policy which emphasise the importance of the government's reputation, the vicious circle can take on an additional twist. Here the use of devaluation damages a government's anti-inflation credentials; private agents anticipate devaluation and mark up prices ahead of it; the inflation thereby caused itself forces the government to devalue. Expectations become self-fulfilling and generate their own internal dynamics.

The Fund has also been seen as being over-ambitious. Its stabilisation and liberalisation objectives have been interpreted as paying inadequate regard to the potential inconsistencies that may exist between them. Within developing countries, in particular, revenue from tariffs may be an important element in total government income. Tariff reduction can therefore exert a significant adverse impact on the fiscal balance unless this source of revenue is replenished by other tax changes.

Evidence suggesting a falling rate of success in achieving programme targets is cited as supporting the claim that Fund-supported programmes in highly indebted countries have been unrealistic.[12] In the case of intermediate targets, relating, for example, to aspects of credit creation, such a record reflects an increasing problem of non-compliance. Countries have often simply not complied with strategic elements in Fund-supported programmes. Some authors have again sought to explain this phenomenon in terms of the specifics of the debt problems with which highly indebted countries have been faced, the argument being that Fund-supported programmes have offered little *domestic* rate of return. The principal beneficiaries have instead been private foreign creditors. The distribution of the costs and benefits of the programmes has established a set of incentives that is antagonistic towards a high degree of compliance. The debt overhang has had the effect of weakening Fund conditionality through acting as a tax on necessary reforms, with one implication being that it has become increasingly difficult to muster the necessary domestic political support for such reforms (Sachs, 1989b; Krugman, 1988). In this context it is claimed that debt relief is needed to create the necessary incentive structure to adjust. The Fund has been criticised for failing to recognise this. Indeed, its policy of 'assured financing', whereby IMF support was predicated on countries continuing to meet their outstanding obligations to the banks, has been interpreted as systemically discouraging the provision of debt relief by the banks and thereby impeding the resolution of the debt crisis. At the beginning of the crisis the Fund had some success in encouraging new commercial money inflows by making these a precondition of its support, but this insistence faltered as the banks' reluctance to lend became more pronounced.[13]

Moreover, it is argued that the Fund's inappropriate approach to the debt problem was reflected by its apparent neglect of the distinction between new financing and debt reduction – a distinction which was being accentuated in the academic literature as the 1980s progressed (Krugman, 1988). Critics suggested that this neglect again showed the Fund as being primarily concerned with cash flow rather than medium and longer-term problems. Yet, even in a short-run context, the different expectational responses to new money and debt reduction can cause different effects, with new money leading to further indebtedness and therefore the prospects of additional domestic fiscal and monetary problems.

Statements emanating from the Fund about its own perception of its role in the debt crisis tended to side-step these analytical issues and stick with broader organisational ones, which emphasised its strategic importance as an 'honest broker' or catalyst (Nowzad, 1989). The Fund described its objective as that of normalising creditor–debtor relations and restoring country access to sustainable flows and spontaneous lending. The means to this end were to be vigorous and sustained adjustment efforts by the debtors, and a co-operative concerted approach involving creditors, the Paris Club, commercial banks and the export credit agencies. While recognising that progress had been uneven and vulnerable, by the mid-1980s the Fund was interpreting its overall record on the debt problem as 'encouraging' (Nowzad, 1989). At the same time, however, critics were assessing that, 'the IMF's recent record in the debtor countries is one of failure' (Sachs, 1989a).

Such disagreement persists because there is no universally accepted set of criteria by which the Fund may be judged. Apart from anything else, there is always the basic problem of the counterfactual: what would have happened if the Fund had done things differently? Accepting this difficulty, a superficial review of the empirical evidence suggests that the Fund's record in terms of dealing with the debt problem of the 1980s was, at best, mixed. Certainly it managed to help avoid a major systemic international financial failure and this was no small achievement. But, by other criteria, no substantial or sustained degree of success can be claimed. By the end of the decade, creditor–debtor relations had not been normalised, and access to spontaneous lending had not been restored. Indeed, the creditworthiness of the highly indebted countries, as represented by the secondary market price of their debt, had continued to fall; net transfers to highly indebted countries were still significantly negative; a concerted and co-operative approach to the debt problem had not emerged; most debt indicators failed to show any notable or sustained improvement; and macroeconomic performance in the highly indebted countries was poor and often deteriorating, with forward-looking indicators such as the investment ratio and import volume suggesting bleak prospects for the 1990s. Even IMF-specific indicators were discouraging, with declining programme compliance, rising arrears and the increasing use of waivers. Episodic successes existed but the overall picture was not reassuring.

During a decade in which open economy macroeconomics became more sophisticated, the accusation was increasingly made that the model underpinning the Fund's operations had failed to be modified and that it was out of date and inappropriate.[14] Research of an excellent academic standard conducted within the Fund's own Research Department was, according to this view, no longer having a significant operational impact. Indeed, and again at a superficial level, the empirical evidence seemed to suggest that the conventional caricature of a Fund-supported programme involving a combination of exchange rate devaluation and the deflation of aggregate demand through credit control was more accurate during the 1980s than it had been before (Edwards, 1989).

At the same time as Fund-supported programmes were being criticised for lacking intellectual sophistication, evidence as to their adverse social and human implications was also being more systematically collected and coherently presented (Cornia *et al.*, 1987; Demery and Addison, 1987). Increasing infant mortality and morbidity, malnutrition and falling life expectancy were now being attributed, at least in part, to IMF-backed programmes. And the design of programmes which emphasised reduced government expenditure rather than increased tax revenue was being seen not only as endangering important welfare schemes in developing countries, but also as reflecting the dominant current politico-economic paradigm within the developed countries, where the role of the state was under stark review. This in turn highlighted another area – the sequencing of reform – in which the Fund came in for criticism. Merely designing an appropriate programme of policies was now seen as inadequate; more consideration needed to be given to the order and inter-temporal distribution of elements of an adjustment programme, particularly as even research conducted within the Fund itself was beginning to suggest that Fund-supported programmes could have a negative effect on output, at least in the short run (Khan *et al.*, 1986; Vines, 1990). Earlier models, which formed the basis for financial programming within the Fund, most notoriously the Polak model, had basically assumed away such an effect by making output exogenous.

Yet even the more outspoken critics of the Fund's handling of the debt crisis suggest that its approach changed towards the end of the 1980s, particularly after Michel Camdessus took over as Managing Director in 1987. This change of approach found expression in terms of a softening attitude towards debt relief, a change in the treatment of arrears, with the Fund becoming prepared to make loans while countries were in arrears with the banks, and an increasing concern for the effects of Fund-supported programmes on income distribution and the related recognition that income distributive effects might be important in determining the political, and therefore practical, feasibility of programmes.[15] Although criticisms still remained, for example that the Fund placed too much reliance on voluntary forms of debt reduction which, given the associated free rider problems, should instead be treated as a public good, they became slightly more muted. If the Fund was still not coming up with right answers, at least, according to some critics, it seemed to be asking more relevant questions. Moreover, some of the broader criticisms relating to the input of the Research Department were suspended awaiting the impact of the appointment of a new Managing Director. On top of this there appeared to be a growing acceptance that macroeconomic stability was a necessary precondition for sustained economic development, and this took some of the sting out of the old debate about IMF conditionality. At the beginning of the 1990s private capital began to return to some of the lightly indebted countries, to the extent that some commentators claimed that the Latin American debt crisis was over. This was not the case in Africa, and it is unclear as to how significant the Fund's input was in generating capital inflows.

## THE FUND AND LOW-INCOME COUNTRIES

If the Fund came in for criticism in the context of its involvement in the highly indebted developing countries during the 1980s, it has sometimes been presented as being almost totally misplaced in the context of the low-income countries. As noted earlier, by the end of 1981 the Fund was almost exclusively involved in such countries and even by the end of the decade the majority of the programmes it was supporting were with them. Many of its staff speak with some regret of how events have altered the focus of the Fund. They view it as being well suited for a broad systemic role and for involvement with economically sophisticated industrial countries suffering temporary payments problems. And yet what they observe is the loss of a systemic role, and little involvement with industrial countries but instead heavy involvement with developing countries which have deeply embedded, if not quasi-permanent, structural balance of payments deficits. The Fund feels organisationally uncomfortable in this role, accepting, or even suggesting, that what is needed in low-income countries may be different from what it can currently provide, given its Articles of Agreement.

Basically, the Fund's view seems to be that what low-income countries frequently need is more foreign aid.

However, the argument can be overplayed. If it is a fact that 'conventional' Fund lending and adjustment programmes are inappropriate for low-income countries then this may imply not only that the Fund is the wrong agency to be involved but also that it needs to show a greater willingness to change its normal operating procedures to accommodate the special needs of these countries. Moreover, while it may be understandable that the Fund's own staff take them as given, the Articles of Agreement can be changed by the Executive Board. In any case, the Fund prefers to present itself as a balance of payments institution, and there can be little doubt that low-income countries encounter balance of payments problems, the size and nature of which were reflected by the data presented earlier as being generally greater than those in other developing countries. Given these problems and the paucity of alternative financing channels, the Fund is unlikely to be able to side-step its involvement in low-income countries, whatever its preference may be. While no one would disagree with the claim that the economic problems facing low-income countries are severe and often appear intractable, failure to search for improvement will do nothing to help.

As has been argued by critics in the case of the highly indebted countries, if the IMF is setting targets that are unachievable, then the programmes it is supporting will be unsuccessful and problems with arrears and ineligibility will follow. Moreover, the Fund will lose credibility and reputation. It is, of course, the loss of reputation associated with unsuccessful programmes in low-income countries that in part explains the Fund's generally unenthusiastic attitude towards lending to them.

But what is the Fund's track record in such lending? Some commentators have suggested that it is likely to be poor. They argue that in low-income countries more than anywhere else it is impossible to draw a distinction between the balance of payments and development. Payments deficits are of a structural type reflecting the fact that these countries have undiversified product mixes, with a high concentration on products that generally have low price and income elasticities. Furthermore, reliance on a few key exports makes them particularly vulnerable to external shocks and movements in the terms of trade. The resolution of these problems is a long-term process which needs to focus on the supply side of the economy. The short-run management of demand is important, but fails to capture the essence of the problem.[16] Moreover, critics suggest that Fund-supported programmes, which rely on changing behaviour by means of altering domestic relative price incentives, will be at their least effective where elasticities are low and where markets are not fully developed. Thus, devaluation has been most commonly criticised in the context of low-income countries. With ill-developed financial markets, fiscal deficits are more likely to be monetised, with consequences for inflation and the balance of payments. Tax revenue may be insensitive to changes in tax policy and government expenditure may be difficult to cut. On top of this, to the extent that Fund-supported programmes have a contractionary effect on the countries that implement them, this may be deemed to be of particular concern in the poorest countries.

The evidence seems to lend some support to these concerns. For a number of years researchers have argued that the elements of adjustment programmes do not appear to discriminate between low-income countries and other borrowers (Killick et al., 1984). At the same time, programmes appear to be relatively less successful in low-income countries than elsewhere. Indeed, some empirical studies have suggested that, even on the basis of methodologies used by the Fund, there are few significant differences to be found between those low-income countries that have adopted Fund-supported programmes and those that have not.[17]

Consistent with the data on the use of Fund credit presented earlier, it also transpires that, once having turned to the Fund for financial assistance, low-income countries find it particularly difficult to disengage themselves. The 'league table' relating to the number of consecutive years over which Fund credit has been outstanding is dominated by the low-income countries of Africa. The reality in such countries is therefore at odds with the idea of the IMF as an institution offering temporary balance of payments assistance.[18]

Arrears to the Fund, while confined to a limited number of countries, have been more of a problem for low-income countries than for others.[19] This has symptomised the fragility of their balance of payments. This fragility, scant access to private sources of financing and their small cushion of international reserves have pushed low-income countries more heavily to the Fund than other groups. When there has been a move within the Fund towards stricter conditionality, this has therefore had a particularly marked effect on such countries, which will, as a result, be more affected by any inappropriateness in the nature of Fund conditionality.

It would, however, be unjust to argue that the Fund has been completely unresponsive to the problems of low-income countries. Although not exempt from criticism, the Extended Fund Facility (EFF), the Trust Fund, the Supplementary Financing Facility, the Structural Adjustment Facility (SAF) and the Enhanced SAF (ESAF), as well as the use of subsidies for low-income countries, may all be quoted as evidence of a response. Moreover, it was in 1986–7 in the case of Bolivia, a low-income country, that the Fund adopted an approach to sovereign debt which has been held up by some observers as a model of the approach which ought to be adopted more generally. In this case the Fund did not insist on a further devaluation two months after the most recent one and did not treat arrears to the banks as a breach of its own performance criteria. With the Fund apparently unwilling to act as their *de facto* debt collector, the banks were forced, so it is claimed, to offer debt relief in the form of a debt buy-back scheme. This, along with the arrears themselves, induced and effectively financed the necessary adjustment in the Bolivian economy. In this particular case, it is claimed not only that the Fund-supported programme was successful, but also that the role of the Fund within the economy was politically accepted because it could not be depicted as an agent of the banks, but rather as a net supplier of resources.[20]

By the beginning of the 1990s, the outlook for low-income countries showed few, if any, signs of improving – an outlook hardly helped in the early 1990s by the Gulf War and recession amongst industrial countries. Their balance of payments problems remained entrenched; their creditworthiness in the eyes of the commercial sector remained very low, if not non-existent, as shown by the price of their debt in the secondary market; their need for financial inflows in order to maintain any form of development remained positive, but they had negative net transfers with the Fund. With increasing claims on Fund resources coming from the emerging economies of Eastern Europe, which are themselves unlikely to enjoy much access to commercial bank lending, continuing demands coming from the middle-income developing countries, and a persistent US current account deficit, there is a danger of low-income countries being crowded out in the competition for balance of payments and development financing.

How should the Fund respond to this scenario? To the extent

that low-income countries need more foreign aid, should the Fund explore ways in which lending to them may be made longer-term, more supply-related and more concessionary? Should it explore ways of extending and liberalising those facilities of particular relevance to them, such as the EFF, the CCFF and the ESAF; should it expand the use of subsidies; and should it resuscitate the idea of an SDR aid link? Should it introduce a soft lending window along similar lines to the World Bank's soft loan affiliate, the International Development Association (IDA)? Or should it instead endeavour to disentangle itself from lending to low-income countries by essentially passing the responsibility over to the World Bank or the aid agencies?

## THE FUND'S RELATIONS WITH OTHER INSTITUTIONS

While there is some overlap between the Fund and the various groupings of industrial countries such as the OECD, the G-7 and the G-3 in the pursuit of certain systemic functions including the management of exchange rates and the international co-ordination of macroeconomic policy, it has been the Fund's involvement in developing countries during the 1980s that has brought the question of the institutional division of labour into sharper focus.

The first interface is between the Fund and the commercial banks. Earlier notions had been of the Fund fulfilling a catalytic role and, through its conditionality and its so-called seal of approval, encouraging the banks to lend. Although this has sometimes assumed the proportions of a stylised fact, the empirical evidence of its existence is rather mixed.[21] Indeed, as already noted, by the beginning of the 1980s, the Fund and the banks were essentially involved in different groups of developing countries; a *de facto* division of labour seemed to have emerged, with the banks financing the middle-income countries and the Fund financing the low-income countries. After 1982 this pattern of lending changed dramatically, with the Fund being pulled into lending to the highly indebted countries as the banks sought to withdraw. In the early years of the debt crisis, the Fund essentially acted to impede the banks' withdrawal through concerted lending; it made its support conditional upon the banks continuing to provide finance. To the extent that there was a multilateral debt strategy, it relied on refinancing and new money. The perception within the banks themselves at this time was that they were being asked to do too much relative to the official sector, and they were critical of the Fund. As the 1980s proceeded, concerted lending faltered and the Fund became more exposed to criticisms of a different nature. The accusation now was that the Fund was bailing out the banks and systemically discouraging the provision of the relief needed to resolve the debt problem. Certainly by the end of the 1980s, the banks had become less vulnerable to problems arising from their exposure in developing countries. The international banking collapse threatened by the debt crisis had been averted. Moreover, the emphasis of international policy switched from the new lending associated with the Baker Plan to the debt reduction associated with the Brady Plan. On a small and piecemeal scale, the Fund was now seen as fulfilling the sorts of functions that some observers had suggested should be conducted on a large scale by a new International Debt Facility (IDF).[22]

If the division of labour between the Fund and the banks became fuzzy in terms of country involvement during the 1980s, fuzziness also existed in terms of function. At first sight there seemed to be an argument for exploiting areas of comparative advantage. This would have had the banks mobilising finance and the Fund collecting and processing information and making an input into the design of stabilisation programmes. In fact, the banks themselves began to collect and collate data more systematically through their own Institute of International Finance in Washington, and the Fund, at least at the beginning of the 1980s, took on a financing role.

With the banks endeavouring to minimise their involvement in developing countries, the interface between the Fund and the banks may be of less significance as the 1990s proceed.

Of growing importance, on the other hand, will be relations between the Fund and the World Bank. The division of labour between the two Bretton Woods institutions which had begun to alter in the 1970s underwent still more fundamental change in the 1980s. Prior to the breakdown of the Bretton Woods system, the division of labour had been relatively straightforward. The Fund's orientation was towards the short run: the balance of payments, the demand side, the monetary sector, and programme support. The Bank's was towards the long run: economic development, the supply side, the real sector, and project support. The differences between the two institutions were nicely encapsulated in Keynes' observation that the Board of the Fund should comprise 'cautious bankers' whereas that of the Bank should comprise 'imaginative expansionists' (Moggridge, 1980). Although the line between the balance of payments and development has never been an easy one to draw, for as long as the Fund was not heavily involved in developing countries, this did not constitute a significant problem. The Fund was largely successful in seeking to retain its image as a monetary institution.

In 1966, an internal memorandum clarified the division of labour by assigning 'primary responsibilities' to each agency (Mason and Asher, 1973). The Fund had jurisdiction 'for exchange rates and restrictive systems, for adjustment of temporary balance of payments disequilibria and for evaluating and assisting members to work out stabilisation programmes as a sound basis for economic advice'. The Bank's primary responsibility, in contrast, was 'for the composition and appropriateness of development programmes and project evaluation, including development priorities'.

Less than a decade later, it had begun to be accepted, even within the Fund, that payments deficits could be of a structural nature which required longer-term financial support to correct. The Extended Fund Facility (EFF) was introduced to fill what was perceived as a gap in the range of the Fund's lending facilities. Although researchers have observed that EFF programmes were little different in practice from normal stand-by programmes, and although the Fund's staff were rather unenthusiastic about the facility, its introduction clouded the distinction between the Fund and the Bank. The distinction was further blurred throughout the 1980s, first by the Bank's initiation of a programme of structural adjustment lending through structural adjustment loans (SALs) and sectoral adjustment loans (SECALs), which incorporated conditionality linked to the structural causation of balance of payments deficits, and second by the Fund's introduction of the Structural Adjustment Facility and then the Enhanced Structural Adjustment Facility. In terminology as well as in areas of involvement, structural adjustment had served to create an important area of overlap between the Fund and the Bank.

The agencies themselves attempted to deal with these overlapping responsibilities by seeking to achieve greater co-operation and collaboration and, through this, consistency. The co-operation has been both formal, as incorporated in the mutual design of a policy framework paper (PFP) as part of the SAF, and informal, relying, as some observers have suggested, on the 'personal chemistry' of the relevant staff members. Clearly the overlap can bring with it both advantages, in terms of better informed analysis and judgement, and disadvantages in terms of inefficiency caused by increasing labour input per unit of output, delays and institutional conflict.

Given that there is little or no indication that the structural problems of developing countries will be resolved quickly, the overlapping responsibilities between the Fund and the Bank seem likely to endure throughout the 1990s and beyond, even though the Bank has signalled a partial retreat from structural adjustment lending. Indeed, the overlap will take on a wider geographical connotation with continuing economic reform in Eastern Europe. The question therefore arises: what is the organisational structure best suited to capture the advantages and avoid the disadvantages? Should the Fund and the Bank be merged? Should their areas of responsibility be reasserted or redefined, as seemed to happen in early 1989? Should a new agency be established? Should methods of collaboration be reviewed and reformed? What should be the position on cross-conditionality? Although these may be unattractive issues to economists who often prefer to spend their time designing and testing economic models, they remain important questions from the point of view of those people whose lives are touched by the activities of the Fund and the Bank. Events during the 1980s and 1990s have rekindled interest in this area of international political economy.

Clearly to the extent that institutional reform improved the effectiveness and success of adjustment programmes, this would enhance their catalytic effect on both private loans, if not in the form of bank loans then perhaps in the form of foreign direct investment, and aid flows, where conditionality has been of increasing significance.

## CONCLUDING REMARKS

From being the linchpin of the Bretton Woods system, the Fund was left with an ill-defined systemic role after the collapse of that system. International monetary arrangements began to rely much more heavily on private markets both for foreign exchange and for international capital; the Fund became marginalised. As its systemic role was downgraded and as the industrial countries, and some of the more creditworthy developing countries, turned elsewhere for finance, the Fund was left to deal with the low-income countries that had nowhere else to go. In the 1980s, however, a different sub-group of developing countries was forced to turn to the Fund, which, as a result, became involved in the debt crisis. Its handling of developing country debt has been severely criticised from some quarters and, on top of this, its involvement in the low-income countries has not been a happy one.

Where does the Fund go from here? Does the move back towards greater exchange rate management and international macroeconomic policy co-ordination indicate the possible re-instatement of a systemic role? Following on from the problems encountered by the Uruguay Round of trade negotiations one scenario would, for instance, have the Fund playing a central role in avoiding the protectionist tensions that are seen as being related to the emergence of a tri-polar world economy based on the US, Japanese and European economies. The avoidance of economic conflict between these centres of power, as the hegemony of the US continues to wane and as economic competition replaces military competition, has been seen by some observers as the main challenge facing the IMF. Such a scenario sees the Big Three economic powers acting as a 'steering committee for the world economy' and pledging themselves to the maintenance of a stable international economic order, an aspect of which would be the construction of a new regime to replace the Bretton Woods system, based on target zones and co-ordinated macroeconomic policies. According to this view, 'the IMF would play the critical "honest broker" role in providing forecasts, analyses and policy recommendations to guide implementation of the system' (Bergsten, 1990).

Another related view sees establishing US balance of payments viability as the most pressing current issue in international finance (Finch, 1989). According to this perception, the IMF should seek to re-establish its role in industrial countries where its conditionality could be of significance in restoring confidence. The Fund as a multilateral source of financing is seen as having considerable advantages over bilateral private financing.

An expanded role for the Fund in financing the deficits of industrial countries as well as those of Eastern European countries may, however, have implications for developing countries. Unless the finance comes from sources where they are not competing, or from an expansion in the Fund's lending capacity, developing countries could be crowded. The hard facts of the case, however, are that industrial countries are always going to be less likely to be forced into the Fund than are developing countries, since they will always have superior access to alternative means of financing.

But what is the appropriate response to the problems that the Fund has encountered in its dealings with developing countries during the 1980s and 1990s? Should the pattern of reduced net lending, as seen during much of this period, be encouraged? Indeed, should the Fund be closed down as a lending institution altogether? Alternatively, should attempts be made to overcome these problems by reforming its lending policies? If the Fund is closed down, what chance is there of meeting the financing needs which developing countries are likely to encounter during the 1990s? If the market fails is there a role for the Fund to play? Even if a strong case can be made in favour of a role for the Fund during the 1990s, a central issue remains whether its major shareholders will support such a role.

The banks have no desire to renew lending to the levels of the 1970s, and foreign direct investment, although a channel which may be broadened, is unlikely to assume the necessary quantitative proportions. Moreover, it is unlikely that foreign aid will increase sufficiently to fill the financing gap that faces developing countries. The unpalatable alternative is that they will have to reduce their demand for international finance by contractionary and protectionist policies. A lesson of the 1980s, however, is that pursuing this alternative can lead not only to reduced living standards in those countries in which the adjustment occurs, but also to political and social unrest and instability. Because of this the industrial countries may perceive themselves as having a vested interest in reforming the Fund's role in developing countries.

A firmer commitment to the Fund would, of course, bring with it the question of the adequacy of its own resources, and the ways in which these are provided. Should the Fund continue to rely on subscriptions; should it borrow more heavily from some of its members as a means of recycling international finance; or should it borrow from international capital markets? Could it not more actively explore the greater use of the SDR?

The remainder of this book focuses on some of these questions.

Although it is impossible to draw a firm line between the Fund's lending role and its adjustment role, the focus here will be on the former as much as possible, only raising issues of adjustment and conditionality where these have direct implications for Fund lending; adjustment issues are dealt with in detail in the companion volume in this series. It is hoped that the book is timely. While much attention has been paid to IMF conditionality, the Fund as a lending institution has been rather overlooked. This book seeks to redress the balance. To this end Chapter 2 examines the principal analytical issues involved in IMF lending. Chapter 3 undertakes a detailed empirical investigation not only to provide a picture of what the Fund's lending activity actually is, but also to see whether some of the analytical issues may be resolved empirically. The final chapter draws out a range of policy conclusions and discusses the way ahead for lending by the IMF to developing countries.

# 2

# IMF LENDING
## The analytical issues

The IMF may be viewed as both a financing and an adjustment-orientated institution. Through the prolonged debate there has been about Fund conditionality it is the adjustment role that has received the greater scrutiny, but it is important not to lose sight of the financing role. Indeed, an optimal strategy for dealing with balance of payments deficits is always likely to involve a combination of adjustment and financing;[23] and the role of the Fund may be perceived as trying to ensure that this optimal blend is achieved both globally and at the level of individual countries.

However, while the basic rationale of Fund conditionality is rarely challenged nowadays, it has been suggested forcefully by some observers that the Fund should not make loans of its own. Not only is Fund lending viewed as being of no benefit, but it is also seen as being harmful. This chapter assesses the issues raised in this context, examining first of all the general moral hazard case against Fund lending. It then goes on to examine a series of more specific analytical issues associated with Fund lending. Most of the chapter focuses on Fund lending through the General Resources Account and through various additional accounts designed especially for poorer countries, but, towards the end, there is some discussion of the Special Drawing Rights (SDR) account. Historically developing countries have argued that there is substantial scope for enhancing inward resource flows through modifications to the SDR facility. The fact that SDRs have not been created since 1981 has made such proposals appear largely irrelevant. But a rekindling of interest in the SDR could transform the situation and raise again the question of using SDRs as a means of facilitating international resource transfers.

The chapter also briefly discusses alternative ways in which the

Fund's lending operations may be financed, examining the options of quota-based subscriptions, borrowing from members, borrowing from private capital markets, and the use of SDRs and gold.

This chapter should be read in conjunction with the following one, which provides an empirical analysis of Fund lending. This builds on many of the issues raised here, not only analysing the size and pattern of Fund lending inter-temporally and cross-sectionally but also investigating the cost of Fund finance, its composition across the range of Fund facilities, and its significance in relation both to the problems with which it is attempting to deal and to other sources of international finance.

## FUND LENDING: GENERAL ANALYTICAL ISSUES

### 'Hard' and 'soft' critiques of Fund lending

Should the Fund be putting any of its own resources into balance of payments financing? Why not leave such financial provision to private international capital markets? Those approaching these questions from a neo-liberalist angle argue that Fund lending is both inefficient and, to the extent that international equity is relevant to the Fund's operations, also inequitable. The principal source of inefficiency arises from the supposed 'moral hazard' of Fund lending. The suggestion is that the governments of potential borrowing countries are enticed by the offer of balance of payments finance at subsidised and concessionary rates to pursue policies which create the deficits without which they would be ineligible for Fund support. Fund lending thereby contributes to causing the very problems that it is supposed to help solve. Furthermore, it is claimed that the availability of Fund resources makes indebted countries less willing to pursue the adjustment programmes needed to enable them to meet their outstanding debt-service obligations. Fund lending is again portrayed as creating a moral hazard by rewarding those countries that threaten default and demand debt rescheduling or debt relief. Lending by the Fund is seen as undermining its adjustment role.

But why does the IMF continue to lend if the moral hazard argument is legitimate? Two answers are offered. The first is that continuing, and even expanded, lending is what the economic theory of bureaucracy and public choice would predict. Here the

Fund's implicit organisational utility function is seen as comprising its own size, power and influence. According to this line of argument the Fund will try to expand its portfolio of loans as well as progressively to increase the spread and strictness of conditionality. To better achieve its own organisational objectives it will also favour the subsidisation of its loans in an attempt to increase the demand for them.[24]

The second answer is that the Fund believes, mistakenly according to neo-liberalist critics, that the benefits associated with its lending outweigh the costs expressed in terms of moral hazard. Vaubel (1983), for example, presents and then attempts to dismiss various arguments for Fund lending. First, the defence of exchange rates is argued to be inappropriate in the context of a generalised flexible exchange rate regime; and, in any case, IMF lending as a means of financing the intervention required to defend exchange rates is argued to be unnecessary even with fixed exchange rates, since such intervention may be substituted by sufficiently restrictive domestic monetary policy. Second, the selection of more gradualist adjustment programmes permitted by Fund lending presupposes their superiority over the alternative shock treatment. Shock measures may, however, be the ones that are required. Where gradual adjustment is appropriate, neo-liberalist critics of Fund lending argue that this will be supported by loans from private international capital markets. Again, according to this view, the market should be left to determine the international distribution of capital and the Fund should not become involved.[25]

Third, the insurance argument, namely that the Fund is needed as a lender of last resort, is rejected on the grounds that the international banking system is much less susceptible to collapse than is often supposed; that banks making unwise loans should not be protected; that the insurance role may be played by the monetary authorities of the developed countries; and that, in practice, the Fund is frequently used as a first rather than as a final line of credit, a use that the Fund itself encourages by offering concessionary loans.[26]

A fourth argument, that Fund lending discourages recipients from pursuing undesirable beggar-my-neighbour policies and therefore confers positive externalities, is rejected on the grounds that the Fund should not attempt to buy 'social peace' by giving in to 'blackmail'. Such a policy is presented as myopic on the grounds that it encourages further threats and ultimately aggravates international discord. The more benign 'mutuality of interests' motivation for financial assistance is rejected as being an inefficient means of correcting deficiencies of aggregate demand in donor countries.[27] A higher multiplier applies to domestic government expenditure.

Fifth, critics argue that Fund lending is not needed in order to encourage countries to pursue economic adjustment, since creditworthiness in the international capital market will provide sufficient incentive. Vaubel (1983: 298) claims that 'in the market, conditionality is automatic, perfect and unavoidable'. Although the pursuit of an appropriate adjustment programme will justify a reduction in the rate of interest that a borrowing country has to pay, because of reduced risks of default, it is not seen as justifying a relatively lower rate on Fund lending than on commercial loans.

'Enforcement', 'bogeyman' and 'coherence' arguments claim that the Fund is better placed to enforce conditions through the imposition of sanctions; that governments need someone to blame for unpopular policies; and that a co-ordinating agency representing creditors is needed in order to design coherent policies. These arguments are interpreted as potential reasons for the Fund's existence as an adjustment institution but not as justification for a lending role. Similarly the argument that the Fund possesses superior information is, if accepted, seen as an argument for more freedom of information rather than as an argument for Fund lending. Information, it is claimed, is a public good.

The use of subsidies on Fund lending, either explicit in the case of the ESAF or implicit under other lending facilities in the form of a discount by comparison with commercial rates, is presented as being inefficient. It encourages borrowers to delay adjustment, provides an actuarily unfair insurance benefit, and leads to problems of adverse selection in the sense that the largest subsidies, measured by the difference between Fund rates (which are uniform) and commercial rates (which will vary according to perceived risk), are offered to the least creditworthy countries which, so it is claimed, have the most negligent governments.

A rather softer critique of Fund lending does not go so far as to suggest that it should be discontinued altogether. However, the implicit notion here is of an optimum quantity of Fund lending that can be in principle, and has been in practice, exceeded. Fund lending will be excessive to the extent that it is linked to inadequate adjustment programmes. The costs of excessive lending by the

Fund will be a loss of its reputation and credibility which will eventually undermine it as an international financial institution. The loss of reputation will erode the catalytic effect that Fund-supported programmes and Fund lending are claimed to have on other financial flows.[28] The result will be that excessive Fund lending will, in the long run, have a net negative impact on the total flow of international finance to those countries that currently receive loans from the Fund.

We return to assess this softer critique of Fund lending a little later. The more extreme criticism essentially comes down to the view that private international capital markets are more efficient in allocating resources than is the IMF. Fund lending is claimed to provide an illustration of government failure.

## Justifying Fund lending: moral hazard or market failure?

Unsurprisingly, the justification for Fund lending rests heavily on the claim that moral hazard is not associated with it and that any government failure is dwarfed by the size of the market failure associated with relying on private flows. Although a defence of Fund lending may therefore be mounted in fairly conventional terms, they are in fact terms that have remained rather resilient to the resurgence of the neo-classical paradigm.[29]

For a number of reasons it is illegitimate to criticise Fund lending on grounds of moral hazard. Evidence to be examined in Chapter 3 suggests that many countries prefer to borrow from private capital markets, albeit at higher interest rates, at times when they would be able to draw from the Fund. This has certainly been the case for all industrial countries over recent years, but it has also been true for a large number of developing countries. Members of the Fund are disinclined to borrow from it because of the conditionality that is attached to the loans. It was, for example, the desire to avoid IMF conditionality that led Latin American countries to borrow from the commercial banks during the 1970s and early 1980s in preference to the Fund. If IMF conditionality is not, and is not perceived as, a soft option, it may reasonably be presumed that conditionality more than offsets any moral hazard associated with the Fund's provision of relatively cheap finance. By the same token, of course, any relaxation in conditionality may enhance the moral hazard criticism. In any case data to be presented in Chapter 3 call into question the assumption that Fund finance is highly and universally concessionary. In particular, for those Fund resources that are financed by borrowing the rate of interest may not be significantly below the market rate, although it has to be noted that potential borrowing countries may face an availability constraint and be unable to borrow at market rates.

In the past not all Fund credit has been subject to strict conditionality. Is the moral hazard problem more relevant in the case of low conditionality lending? In fact, even here it turns out to be largely illusory. The Fund's most significant source of low conditionality finance has been the Compensatory Financing Facility (CFF). Yet members of the Fund have only been eligible for support under this facility where their payments problems have been caused by export shortfalls (or import excesses on purchases of cereals) that are assessed as being beyond their control. The external causation element has therefore neutralised any potential moral hazard. After 1983 the CFF became, and since 1988 the remodelled Compensatory and Contingency Financing Facility (CCFF) has been, a high conditionality facility. The CCFF thus circumvents the moral hazard criticism via the conditionality route. It would therefore seem that at worst the moral hazard criticism of Fund lending applies only to drawings under the first credit tranche and is thus of minimal quantitative significance.

Public choice theorists have attempted to deal with the conditionality counterthrust to moral hazard by arguing that conditionality is *ex post* and will not therefore induce the efforts to avoid balance of payments problems that would be associated with *ex ante* and previously announced conditionality rules. Again the apparent logic of *ex post* (and confidential) conditionality is to maximise the institutional discretion and power of the Fund.

The public choice approach to Fund lending may be extended in a number of ways. First, the assumption is made that the Fund seeks to maximise an objective function which incorporates conditionality and lending. Utility is viewed as a positive function of both, but increased conditionality reduces the demand for loans and the Fund therefore seeks a utility-maximising combination of conditionality and lending. Where there is an exogenous increase in the demand for Fund credits this will result in both increased lending and increased conditionality. Optimisation may, however, be constrained by the Fund's own lending capacity. Here conditionality will tend to be stricter than it would otherwise be, and a relaxation in quota restraints will lead to a reduction in the degree of conditionality as well as an increase in lending.

A second extension of the public choice model is to view IMF lending as the out-turn of a bargaining process between the Fund and borrowing countries. For the borrowers, utility is positively related to the size of credits but negatively related to conditionality. The actual combination of conditionality and lending will depend on the bargaining strength of the two parties. However, the disutility of conditionality from the borrowers' viewpoint will increase and they will then become less willing to opt for packages that involve stricter conditionality unless accompanied by increasing lending (see the Appendix to this chapter for a more formal presentation and critical discussion of the model).

If this type of model is accurate one would expect to see significant variations in conditionality not only over time, as the demand for Fund credits and the lending capacity of the Fund change, but also across countries reflecting differences in bargaining strength. Changes in conditionality would themselves, of course, induce changes in demand for Fund credit. To some extent the model may be assessed against the empirical evidence, as is done in more detail in the following chapter. However, the basic assumption that the Fund seeks to maximise its power and influence is either at worst a gross contortion of its rationale and *modus operandi* or at best an oversimplification which does not allow for a complex decision-making process which involves the management of the Fund and its Executive Directors. An alternative picture is that the Fund is seeking, in an ongoing fashion, to modify conditionality in ways that maximise its positive effects on the borrower's overall balance of payments. As part of this process of evolution a relaxation in conditionality may coincide with an increase in lending, or a tightening up may reduce the demand for loans; but this does not necessarily confirm the public choice model. Even endeavouring to manage demand via conditionality may be seen as a legitimate aspect of the Fund's role as an international financial institution.

While studies have discovered some variations in IMF conditionality across countries and over time (Killick *et al.*, 1984; Killick, 1992) the hard core of Fund-supported programmes appears to be time invariant and is well known by potential borrowers *ex ante*. Nor is it true that failure to comply with performance criteria carries no penalty. Although deviations may in some cases be waived, in others they will result in a discontinuation of lending. Moreover, countries may believe that failure to meet conditions will undermine their future access to finance both from the Fund and perhaps more importantly, elsewhere. Indeed, the costs of so-called slippage or non-compliance are a significant factor in the moral hazard analysis of Fund lending. For as long as governments of borrowing countries believe that non-compliance will damage current and future financial flows, and to the extent that they are risk-averse, they will not take conditionality lightly. Of course, where slippage is seen as carrying no penalty the Fund loses a potential 'stick' and in this case the structure of incentives will not be as strong as it might be and the moral hazard argument has more validity.

However, the moral hazard critique of Fund lending stands or falls on the assumption that governments continue to create balance of payments circumstances in order to gain access to Fund finance. Indicators of domestic economic mismanagement have indeed been found to exist in many of the countries drawing from the Fund. But it is a large and illegitimate step to move from this observation to the moral hazard argument. First, it cannot be assumed from this that governments set out to create exchange rate misalignment, fiscal deficits and excess money. Failure to achieve what would be widely recognised as reasonable economic objectives does not necessarily imply success in optimising a *different* set of objectives. Second, if the incentives to borrow from the Fund are so strong, why do all countries not set out to create balance of payments deficits? If the moral hazard argument is powerful it would be reflected by the Fund always being up against its lending capacity.

Finally, and significantly, there is an accumulating body of evidence to suggest that balance of payments deficits are caused not only by domestic economic mismanagement, but also, and importantly, by adverse terms of trade movements and increasing real interest rates.[30] Again the moral hazard critique of Fund lending is undermined.

When it comes to international lending, the claim that the private capital markets know better than the Fund has been exposed to considerable criticism over the past ten years, particularly in the context of private bank lending to developing countries. In response to the question 'why not leave international financial provision exclusively to the market' the short conventional answer is that the market mechanism is deficient on grounds of both efficiency and equity. Commercial lending has been characterised by instability and uncertainty, by contagion and bandwagon effects, and by a revealed inability accurately to assess creditworthiness. If private creditors had the perfect foresight that they are sometimes claimed to possess the developing country debt crisis would not have happened. Indeed, a reasonable case may be made that if greater reliance had been placed on the IMF as a source of balance of payments financing during the 1970s and early 1980s, the debt crisis would have been avoided.

Lending by the commercial banks will be inefficient if it is unstable; if it is unrelated to the underlying productivity of resources throughout the world; and if it brings external costs with it. It is certainly unstable. The banks moved rapidly into balance of payments financing following the rise in the price of oil in 1973–4 and then endeavoured to extricate themselves in the 1980s. In general terms they were quick to lend to developing countries following the (temporary) upsurge in commodity prices, and anxious to reduce their lending as borrowers encountered declining terms of trade and debt difficulties. Such instability is enhanced by the tendency towards 'herd behaviour' that characterises bank lending. Withdrawal by one bank can quickly encourage others to follow suit rather than to offset the withdrawal by lending more themselves, or at least maintaining the level of their involvement. Indeed, there is a problem in the sense that all banks will have an incentive to reduce exposure at the first suggestion of repayment problems, yet not all of them will be able to do so, since such behaviour would certainly induce default. As a result of these characteristics bank lending is pro-cyclical rather than counter-cyclical.

As the above discussion implies, lending may be only loosely related to the underlying strength of an economy or the marginal productivity of resources. Banks may have imperfect information and may misinterpret what information they do have; they may be unduly influenced by transient and often largely cosmetic factors or, in syndicated loans, by the views and prestige of the 'lead' bank. This can, of course, work both ways. On some occasions banks may overlend, yet on others they may underlend. Either way, they are unlikely to allocate capital efficiently.

Furthermore, factors that alter the creditworthiness of one borrower can have an influence on the willingness of banks to lend to other countries, which is quite unrelated to their economic performance and prospects. For example, a fall in the price of oil will create debt problems for oil producers such as Mexico and Venezuela and will damage their credit rating with the banks. But as a result of these debt problems the banks may become more risk-averse and less prepared to lend to other countries whose economic prospects actually improve as oil prices fall, because the price of a major import has fallen and because a falling oil price tends to increase world aggregate demand and therefore the demand for their exports. The lack of simultaneity in the debt-servicing capacity of borrowers suggests that such a response is irrational, though to the extent that the withdrawal of funds itself creates a liquidity and perhaps, with rising interest rates, even a solvency problem in the affected countries, it may seem rational after the event.

Another externality associated with bank lending relates to its global consequences. Where banks pull out from providing balance of payments finance and nobody else steps in to take their place, the result is that borrowers have to correct their deficits more rapidly. Rapid correction can be brought about by deflating domestic aggregate demand, and thus the demand for imports, or by introducing import controls. Either way, the countries providing the imports will experience a reduction in their exports and a deterioration in their balance of payments, which they, in turn, may have to correct. A vicious circle of deflation, recession, protectionism and falling world trade can become entrenched, from which most countries stand to lose. Furthermore, the mere uncertainty of bank lending may give rise to external costs.

But, at the same time, is it really the banks' responsibility to ensure that things do not go wrong in this way? They will undoubtedly see their principal responsibility as being to their shareholders, and they are likely to take the view that these interests are best served in an uncertain world environment by trying to maximise short-run private profits. It may be unreasonable to expect them to assume the global role of maximising world economic welfare.

The concept of welfare also raises the question of the distribution of bank lending. In aggregate terms this has been heavily skewed, being concentrated in industrial countries and a narrow range of middle-income developing countries. The banks have usually deemed low-income countries uncreditworthy.

Again, however, it is unreasonable to criticise the banks for the inequitable distribution of their lending. After all, they are not charitable institutions. Indeed, on the contrary, they would be more open to criticism were they to lend to countries that seemed to have little chance of repaying the loans. Yet, the elements of market failure remain, not only with respect to equity but also in terms of efficiency. If lending by the banks fails to meet the requirements of an efficient and equitable market solution, should the recycling of world capital be left in their hands or should international agencies such as the Fund not become involved? The Fund is also better positioned than the banks to take a global view of economic welfare.

Further doubts about the wisdom of relying too heavily on private bank lending are raised in a measured assessment of the international capital transfer process by Llewellyn (1990) which draws heavily on earlier work by Lessard and Williamson (1985). He identifies the main requirements of an ideal process as follows (Llewellyn, 1990: 35):

1 The mechanisms should facilitate (or at least not impede) an effective selection of projects such as to maximise the risk-adjusted rate of return on world capital.

2 Institutions and markets would offer a wide range of instruments and financing facilities with respect to maturity, risk, rate of return, contract formulae, etc., so that a balanced portfolio of liabilities is available to borrowing countries. A balanced portfolio is required to avoid concentration on particular types of servicing contract, and a dependence on a single source of finance whose supply could be interrupted.

3 The supply of finance should be reasonably stable and not subject to large discontinuities.

4 The nature of the contracts would offer a reasonable match between debt-servicing obligations and the borrower's ability to pay, both with respect to maturity and cash flow. Lessard and Williamson (1985) identify 'cash flow matching' via, for instance, risk-sharing equity contracts, hedging instruments, etc., as a major part of their proposals for improving upon the quality of finance in future arrangements for financing the net absorption of real resources by developing countries.

5 The mechanisms should allow for the diversification, transfer and sharing of risk to where they can be most effectively

absorbed because of the holder's capacity to reduce risk via diversification.

6  There would be an efficient system for identifying and pricing risks on a continuous basis.

7  Lessard and Williamson also identify the issue of performance incentives whereby the intermediation mechanisms would create the correct incentives for the performance of the borrower.

8  A displacement of domestic savings and local financial markets should be avoided. Lessard and Williamson suggest that this did happen in the case of some heavily indebted countries through the route of 'capital flight' due to inappropriate policies by governments. They suggest that general balance of payments financing from the international banking sector enabled governments to bypass local markets (international intermediation displaced domestic mechanisms).

9  International financial intermediation mechanisms would limit systemic hazards for financial systems supplying financial intermediation services.

10  There should be no incentive to repudiate debt.

Llewellyn goes on to claim that on the basis of almost every one of these criteria banks were an inappropriate conduit for international resource flows during the 1970s and early 1980s, and remain so in the 1990s.

There are, of course, ways in which private bank lending may be made more efficient. If methods of country-risk analysis have been inadequate in the past, they can be improved.[31] The results of such improvements would be to make bank lending first more stable and predictable, second more closely related to significant economic factors determining the future ability to service debt, and third less subject to the regionalisation phenomenon. In practice, however, banks have tended to respond to the developing country debt crisis not by putting their resources and efforts into making risk analysis more sophisticated but rather by simply taking decisions not to lend to developing countries. Refining the techniques of risk analysis has therefore been seen as decreasingly rather than increasingly important. It is, moreover, difficult to see that the resurgence of bank lending to some developing countries in the early 1990s which was reported in Chapter 1 is based on any superior assessment of risks.

The observation that the banks possess less information than the Fund and therefore will always tend to reach inferior decisions cannot legitimately be dismissed by arguing, as some have attempted to argue, that this only reflects a prejudicial unwillingness to accept the judgement of the market. The Fund does possess superior information. Although the Institute for International Finance collects and processes information on behalf of the banks, the Fund remains in a uniquely strong position to collect and interpret information and to act on the basis of it. From the viewpoint of political economy it would be exceedingly difficult to make all this information openly available to banks. If countries were to lose the confidentiality that they enjoy with the Fund, the flow of accurate information would be reduced, and this would make the capital transfer mechanism yet more imperfect.

Some of the deficiencies of bank lending may be overcome, in principle, by shifting to other forms of private lending such as bonds, or indeed foreign direct investment as is implied in points 2, 4 and 5 in the list quoted above. There can be little doubt that short-term bank credits are likely to be an inappropriate means of financing economic development or long-term balance of payments problems. But although there is scope for improving the financial instruments through which private capital is internationally made available, it remains unlikely that the ultimate lending decisions will be efficient. Private capital markets are not well equipped to evaluate risk where lending is programme-based.

To what extent, however, is there an unwillingness to accept the judgement of private capital markets because it fails to coincide with some predetermined preferred result? Is this an important factor in the market failure debate?

Although it is quite possible that a particular view of international equity will underpin someone's attitude to the market solution, this is surely perfectly legitimate. The fact that the distribution of private lending fails to coincide with a targeted distribution of financial flows is after all a central justification for foreign aid. But the issue is more complex than this. Without international lending agencies, countries deemed uncreditworthy by international capital markets will be forced to pursue adjustment policies that seek rapidly to correct balance of payments deficits. If it may be shown that such policies impose higher welfare costs than alternative adjustment strategies, then it is not unreasonable to suggest that the international capital transfer mechanism should seek to support these alternative strategies. If, for example, rapid adjustment based on compressing domestic aggregate demand implies shifting an economy inside its production possibility frontier, whereas slower adjustment implies shifting the frontier outwards and raising the supply capacity of the economy, it is reasonable to claim that the second adjustment path is more efficient. Where reliance on private markets would exclude this path, it follows that a market imperfection has been identified. It is correct, of course, that the specification of a welfare function will always be a normative issue. But economists should not for this reason shrink from commenting on mechanisms for internationally allocating capital.

It may be noted in this context that the IMF's involvement in low-income countries does not arise from any specific desire to redistribute world income, but rather from the balance of payments problems which these countries encounter. Given the size and nature of these problems, some degree of financing is appropriate on grounds quite apart from equity. One has to have a very narrow definition of efficiency to argue that the unwillingness of private international capital markets to lend to these countries implies that they should concentrate exclusively on short-run payments correction.

A different, although also normatively charged, approach to justifying Fund lending proceeds by endeavouring to identify financing gaps. To the extent that the gaps are not filled by private markets, a lending role for the Fund – or some other international financial institution – is defined. The difficulty here is again that the gaps only exist to the extent that the actual or predicted state of affairs differs from a desired state. *Ex ante* financing gaps may, in principle, be closed by movements on either the demand side or the supply side. An excess demand for foreign exchange may be eliminated by compressing imports as well as by raising the supply of foreign exchange by international borrowing. If one is indifferent as between the alternative ways of closing *ex ante* financing gaps, then little sympathy for the entire concept of financing gaps will be shown. If, on the other hand, it is felt that the welfare consequences of the alternatives are significantly different, and that the welfare consequences of reducing the demand for foreign exchange are high, there will be a greater desire to spell out more precisely what the financing gaps are.[32]

Starting from a specific target of (say) preventing a fall in living standards, it will be possible to project the likely demand for foreign exchange. Projections of the supply of foreign exchange through export earnings and private borrowing will then identify the size of any resultant foreign exchange gap. The rationale of Fund lending might then be to contribute towards closing this gap. However, the need for Fund lending could easily be removed by abandoning the original specified target. In many ways it is one's normative judgement concerning the desirability of these alternative responses that will affect one's willingness to rely exclusively on private markets as a means of allocating international capital.

The question of what is the 'correct' balance between adjustment and financing is difficult if not impossible to answer with any scientific precision. Where, however, an argument is made for endeavouring to make adjustment the independent variable and financing the dependent one, rather than have adjustment adapt to the availability of finance from private capital markets, an argument is also effectively being made for agencies such as the Fund to be involved in the allocation of capital as lending institutions.

Another way of saying approximately the same thing would be that, if there is little discrepancy between the private costs and benefits of international lending and the social costs and benefits, then it is quite efficient to leave things in the hands of the private markets. However, where the discrepancy is large, the existence of an international agency which takes such externalities into account may be justified.

Let us now return to the softer critique of Fund lending that was raised earlier and that concentrates on the dangers of excessive lending by the IMF. Two observations may be made about this. First, the criticism is one that relates more to the design of IMF-supported adjustment programmes than to IMF lending, although tightening conditionality may consequentially lead to reduced lending by the Fund as the demand for Fund resources falls. Second, and leading on from this, stricter conditionality might, however, generate greater international financial inflows from other sources and enhance the Fund's catalytic effect. The argument can again come close to the claim that the Fund does not need to lend itself but merely to *support* private capital markets via its conditionality.

Although, in principle, offering a way in which the Fund might attempt to deal with a source of market failure, the catalytic effect of Fund involvement has a rather doubtful theoretical pedigree.

Why should the negotiation of a loan from the Fund induce additional private capital inflows? Asset market models of the balance of payments suggest that, to the extent that Fund-backed programmes are associated with a depreciation in the exchange rate and an increase in interest rates, then, other things being constant, a strengthening in the capital account might be expected. However, for countries some distance from the margin of creditworthiness the size of risk premiums may overwhelm such changes. Here, inflows from the Fund may merely finance repayments of private capital. Moreover, the effect of Fund involvement on the confidence of potential investors will depend crucially upon whether there is a general presumption that the negotiated programme will be implemented and will generate beneficial effects. In the absence of such expectations there is little reason to see why the catalytic effect will be positive. Indeed, to the contrary, a history of programme failure and long-term involvement with the Fund may reinforce the view that Fund involvement is a clear indicator of current problems and a lead indicator of future ones. In these circumstances the catalytic effect would, in principle, be negative.

Against such theoretical ambiguity the significance of the Fund's catalytic effect becomes an essentially empirical issue, and is investigated in the following chapter. However, the existence or non-existence of the catalytic effect raises important analytical issues for Fund lending. If it were found to be empirically insignificant the implications would be twofold. First, how can it be strengthened? But second, if Fund involvement cannot be relied upon to generate additional private capital inflows, does the Fund not have to take on a bigger role in helping to close financing gaps by undertaking a larger amount of direct lending itself?

Many of the issues raised so far may be further assessed in the light of empirical information to be presented in Chapter 3. But the general discussion so far has itself raised a number of additional analytical issues which warrant further investigation.

## FUND LENDING: SPECIFIC ANALYTICAL ISSUES

If, as a result of the above discussion, it is believed to be inadequate to rely completely on private international capital markets as a means of organising international resource transfers, a series of additional questions arise:

1 Should it be the Fund or some other international financial institution that makes loans?
2 How large should the Fund's lending role be?
3 Should the Fund's lending role be different in different countries?
4 Through what sort of facilities should loans be organised and how heavily should lending be associated with conditionality?

## The Fund versus other IFIs

The answer to the first of these questions was a good deal clearer ten or fifteen years ago than it is now. If the Fund is characterised as a balance of payments institution and the World Bank as a development one, and if the Fund lends on a programme basis whereas the Bank lends on a project basis, then it would be appropriate for the Fund to be the lending institution where balance of payments difficulties exist and where a programme of macroeconomic policies is required, and inappropriate in other circumstances. But, as described in Chapter 1, a feature of the 1980s and 1990s has been the increasing difficulty in distinguishing between the need for balance of payments finance and the need for development finance. Where does one end and the other begin? To a large extent capital inflows, and particularly those associated with programme loans, are fungible, and it may be unhelpful to try and distinguish between the two types of finance.

Where development finance implies support for specific projects, and balance of payments finance implies support for an adjustment programme designed around reducing aggregate demand, the distinction retains some meaning. But where balance of payments difficulties spring from supply-side weaknesses that will take a number of years to rectify, balance of payments adjustment forms an integral part of economic development. Financial support for a programme that is designed to raise the efficiency of domestic industry, for example, is simultaneously both balance of payments and development finance. This begs a fundamental organisational question. For, if the distinction between the two types of development finance is fairly meaningless in the context of the 1990s, is it equally meaningless to retain a distinction between the Fund and the Bank?

The justification for lending by the Fund, as it is currently constituted, is stronger where a need for short-term credit may be identified. The IMF was, after all, originally designed to provide a revolving pool of credit to member countries rather than as a permanent source of long-term finance. But in many cases, as the data to be presented later will show, the Fund is not fulfilling this function. Rather than lending to a wide range of members for short periods of time it often lends to a relatively small group of countries (given its total membership) over a relatively long period of time. A series of short-term credits effectively becomes long-term financial support. Indeed, longer-term lending by the Fund has been formalised to some extent in the Enhanced Structural Adjustment Facility (ESAF) a facility which also formalises co-operation between the Fund and the Bank.

An initially more clear-cut justification for Fund lending builds on the notion that some Fund members have inadequate international reserves. Reserves and reserve changes are conventionally seen as an indicator of the balance of payments position, although they will clearly also be affected by a country's development strategy. The theory of the demand for international reserves, and the related theory of reserve adequacy, is well established, although it is far from easy to apply in a suitably empirical fashion.[33] Most studies reveal considerable inter-temporal and cross-sectional variation in reserve adequacy. The simple examination of reserve–import ratios in Chapter 1 suggested that some developing countries hold adequate reserves while others have inadequate reserves.

The theory of reserve adequacy implies, however, that where the balance of payments is unstable, adjustment costs are high, and the supply of commercial credit is low, countries will have a relatively high demand for official reserves. Although only Reserve Positions in the Fund count as part of a country's owned reserves, other forms of Fund credit represent conditional international liquidity. Should the Fund be meeting the need for international liquidity?

Further thought reveals that even this approach towards analytically justifying Fund lending is far from straightforward. International reserves are basically an inventory. While they may be decumulated at one stage, they should be replenished at another. The purpose of reserves is to cushion the impact of short-run instability in the balance of payments: they are an inappropriate means of financing a secular balance of payments deterioration or a quasi-permanent deficit; they will eventually run out. Again, therefore, even where IMF lending is perceived as a way of

71

providing conditional reserves to countries experiencing reserve inadequacy, it would be an inappropriate way of permanently financing deficits, unless part of the Fund's role is seen as facilitating such permanent resource transfers; traditionally this has not been the case. Indeed, it has been the belief that the Fund has increasingly been doing precisely this which has led some critics to express concern that Fund lending actually reflects it as a development agency rather than a balance of payments agency.

The general answer to the question of whether the Fund or the Bank should be lending therefore is that it all depends on whether one superimposes a narrow or broad constraint on the purpose of Fund loans. A narrow definition of balance of payments stabilisation and a strict delineation of what constitutes a monetary institution imply that a much smaller proportion of any financing gap should be met by Fund lending; a broader definition of balance of payments adjustment implies a much enhanced lending role for the Fund. It really comes down to a matter of what is the best organisational structure for meeting international financing needs; we return to this issue in Chapter 4.

## The optimum size of Fund lending

The above discussion has also implicitly answered the second question raised at the beginning of this section. The appropriate size of Fund lending depends crucially on the role that the Fund is given to do, as well as on the nature of the international financial regime. Within a regime that emphasises short-term balance of payments adjustment as opposed to financing, and where a significant proportion of balance of payments financing is provided by the private sector, the lending role of the Fund will be relatively small. A modification in regime type towards one which incorporates longer-term adjustment and a smaller financing role for the private sector brings with it an increase in the implied size of Fund lending.

The practical difficulties in translating these general ideas into precise numbers are immense, and are aptly illustrated by the persistent debates there have been over the appropriate size of Fund quotas. What the Fund can lend depends on the resources it has at its disposal, and in large measure these depend on the value of quotas. Thus:

It is generally agreed that the resources available to the Fund should be sufficient for the Fund to play its important role in adjustment and the financing of balance of payments deficits ... While the Fund has had to supplement the resources obtained through quota subscriptions by borrowing from members ... it is generally accepted that its activities should be financed primarily from quota resources.

(IMF, *Annual Report*, 1982: 73)

Yet it has often proved difficult to quantify the Fund's 'important role in adjustment and the financing of balance of payments deficits' in terms of the value of resources it needs to carry out these functions. Bird (1987), for example, presents an analytical framework within which the adequacy of the Fund's 'global quota' may be assessed. The need for Fund resources is thus related to:

1 the size and location of current account balance of payments deficits;
2 the size and distribution of actual and desirable private financial and other financial flows;
3 the size and distribution of non-IMF reserves; and
4 the efficiency and cost of balance of payments adjustment.

However, the model based on these criteria cannot be estimated precisely since many of its elements cannot be objectively measured. The results of the analysis do confirm, however, just how sensitive the Fund's need for resources is to judgements concerning the above factors. The resources made available to the Fund by its principal shareholders therefore implicitly reveal their preferences in terms of the role that it should be performing.[34]

## Cross-country differences

Some of the above discussion also helps us to answer the third question of whether the Fund's lending role should be different in different countries. The theory of the demand for international reserves suggests that demand will be negatively related to the availability of other forms of international liquidity. If loans from the IMF are conceptualised as a form of conditional reserves, then it follows that these will be demanded more heavily by countries which have less access to commercial credit. There is an apparent logic at work here since, if Fund resources are relatively scarce, it will be more efficient to allocate them to countries where the need for them is relatively great. The implication of this logic is that the membership of the Fund may be broken down into various (say three) categories. First, there are the industrial countries where creditworthiness is well established; these countries are unlikely to demand loans from the Fund. Second, there are the low-income countries with persistently poor creditworthiness; these countries are likely to possess an equally persistent demand for resources from the Fund. Third, there are the middle-income countries where creditworthiness varies between countries and over time. Countries in this category will demand resources from the Fund at some times but not at others. This is indeed the pattern of Fund lending that was noted in Chapter 1. Developed countries have not used Fund credit for the last fifteen years or so. Low-income countries, on the other hand, have drawn fairly widely on the Fund throughout the period since the mid-1970s. Middle-income countries tended to avoid the Fund during the 1970s and early 1980s when their creditworthiness enabled them to meet their demand for international liquidity from the private international capital markets, but, as this access evaporated in the wake of the developing country debt crisis, they began to demand loans from the Fund.

Even this simple typology suggests that the basic rationale of Fund lending may be different in the latter two groups. In formerly creditworthy countries, and in those on the margins of credit-worthiness, the rationale may be to restore creditworthiness. For a transitional period Fund and commercial lending may be com-plementary. However, for low-income countries that are a signifi-cant distance away from the margin of creditworthiness, Fund lending is more appropriately seen as a substitute for commercial balance of payments financing.

To the three categories identified above it would be appropriate to add a fourth to cover countries 'in transition' in Eastern and Central Europe. There can be little doubt that these countries will put heavy claims on Fund resources as they struggle to bring about required economic adjustment, and as private markets are reluctant to lend until some signs of achievement are forthcoming.

This typology can, of course, be used by critics of Fund lending, who would no doubt argue that the Fund should not be lending to countries that the market assesses as being uncreditworthy. But even putting this argument to one side, since it sees no lending role for the Fund in any member country, the simple typology of lending does raise some important issues.

If the Fund possesses better information than the market and can therefore more accurately foresee a loss of creditworthiness, should it not be seeking to become involved in countries that at present remain creditworthy in the eyes of the market? Should it not be seeking to prevent a loss of creditworthiness rather than to restore creditworthiness once it has been lost? But then how can countries that are deemed creditworthy by the markets be forced to borrow from the Fund and expose themselves to Fund conditionality?

Moreover, the classification offered above implies that the Fund is a lender of last, and sometimes only, resort. This may be an appropriate systemic role for the Fund, but it does carry with it an important external cost. Countries will only turn to the Fund at a time when no other creditor will lend, or at least will not lend independently of the Fund. At this point balance of payments problems are likely to be deeply entrenched and the required policies will therefore be fundamental rather than superficial. IMF conditionality will be strict. Yet a reputation for strict and far-reaching conditionality may disincline countries from turning to the Fund at an earlier stage in the evolution of their balance of payments problems. The question is how to break out of this vicious and self-perpetuating circle? Success in doing so would, of course, again imply that the Fund would (and should) be lending to countries that may concurrently retain some access to commercial credit.

An interesting empirical issue emerges from this, namely whether Fund lending is used as a complement to or a substitute for private lending. The above analysis suggests somewhat ambiguous conclusions, since the relationship will be different in different countries. On top of this, the credit rating of certain groups of countries will change over time. At certain times the Fund may be used as a substitute for private capital markets but at other times as a complement. Yet ambiguity is consistent with the role of the Fund as an institution designed to compensate for the deficiencies of private markets. In the case where the Fund is acting as a substitute, the deficiency may be expressed in equity terms or in terms of the myopic vision of markets. In the case of complementarity, the deficiency may be of a short-term informational type. It is in this case that a government is attempting to enhance its own credibility by importing the reputation of the Fund in the design of macroeconomic policy. Of course, credibility will be lost where countries are expected to renege on the commitments they have made to the Fund. Maximising the catalytic effect therefore relies on minimising the extent to which countries renege upon or fail to implement programmes negotiated with the Fund. It is in this area of credibility and reputation that pre-commitments to turn to the Fund, and contingency elements in Fund-supported programmes become analytically important. Contingency clauses, for example, which enable Fund-supported programmes to be salvaged in circumstances where they would otherwise have failed will improve the reputation of IMF conditionality and will strengthen the catalytic effect. A pre-commitment to turn to the Fund where circumstances deteriorate will enhance the confidence of private markets and improve current creditworthiness.[35]

## Fund lending: low or high conditionality?

We can now turn to our final question. If there is a case for the IMF to lend should this lending be of a high or low conditionality type? Two key concepts have been used in trying to answer this question.

That emphasised by the Fund has been the distinction between temporary and permanent deficits (Nowzad, 1981). Given the non-transitory nature of most deficits, adjustment is needed. The Fund argues that it is best able to encourage this through the conditions it attaches to its financial support. Left to their own devices, many governments lack political commitment to payments adjustment, and the provision of low conditionality finance by the Fund would therefore merely postpone adjustment with the result that payments performance would deteriorate further, leading to an even more critical need for adjustment.

The second key concept, although seen as being largely irrelevant by the Fund, is the cause or causes of deficits. One view is that the causes of deficits have a vital bearing on the correct balance between low and high conditionality (Dell, 1982). Where deficits are caused by exogenously generated adverse movements in a country's terms of trade the argument is made that strict conditionality is inappropriate. Countries should not be penalised through high conditionality for problems for which they are not responsible.

An intermediate attitude is that, while the causes of deficits cannot legitimately be regarded as insignificant in the conditionality debate, external causation is insufficient reason to attach low conditionality to Fund finance. This view stresses three things. First, that while temporary deficits should be financed rather than corrected, in order to impose minimum cost on economic and social welfare, non-transitory deficits do require correction. Second, that IMF conditionality does have a role to play in encouraging such correction. But third, that conditionality should be appropriate to the economic characteristics of the country concerned, with an important determinant of appropriateness being the cause of the deficits. A central issue here is whether the Fund should become involved with the policies of a country whose payments problems have been externally caused. There is in principle a case for allowing more rather than less discretion to governments that have a good track record of payments management but which have been adversely affected by exogenous factors, and the suggestion has been made that conditionality could be tapered to accommodate various degrees of responsibility (Williamson, 1983).

However, the main problem is to convert these rather vague concepts into measurable and operationally useful ones. There are immense definitional difficulties. How can one distinguish *ex ante* between deficits that are temporary and those that are permanent and therefore between the need for finance and for adjustment? Furthermore, given the problems, on which side is it better to err? Is it possible to state categorically the extent to which any specific payments deficit has been caused by external factors? And, in any case, for what economic phenomena can a government be held responsible? Is it, for example, responsible for a country having a particular level and pattern of production and trade, or merely for controlling the level of aggregate demand? The answer is that it is impossible to be precise about the concepts relevant to the discussion about the balance between low and high conditionality. Some degree of subjective judgement is therefore unavoidable.

On top of this there is again the whole question of the role of the Fund, the role of conditionality, and the correct balance between global adjustment and financing. If the Fund is regarded as an exclusively adjustment-orientated institution, then there is little purpose in having low conditionality facilities if these are simply envisaged as helping to deal with temporary imbalances which do not require adjustment. The private markets might be expected to provide such finance. However, if it is further argued that private markets will be of little use to the least developed countries, then there may be a financing function for the Fund to perform and a related role for low conditionality facilities. Generally speaking, the more the emphasis is placed on adjustment the less will be the relevance of low conditionality facilities. It was indeed different perceptions of the need for adjustment which led to different responses to the first and second oil price increases in the 1970s, with the first being met by an expansion in low conditionality finance and the second by a relative contraction in such finance.

What emerges from this discussion is that there can be a role for low conditionality Fund finance. How important this role is depends on the nature of deficits in terms of their duration and causes, the amount of discretion to be allowed to governments in circumstances where their policies have not been a prime cause of the deficits, and the location of deficits in relation to the creditworthiness of the countries involved.[36]

Although in this chapter we are attempting to retain some distinction between the Fund as a lending and as an adjustment institution, we must reiterate that it is a distinction that is difficult to make at all precisely. The point is that the central feature that differentiates Fund loans from private loans is the conditionality associated with them. Most observers accept that the Fund is in a strong, if not unique, position to design programme-related conditionality and to encourage its implementation. Unless there is compelling evidence of a pronounced catalytic effect, countries will almost certainly need to be offered finance by the Fund as a way of encouraging them to accept its conditionality. Evidence on the use of Fund credit suggests, however, that, even with such finance, countries are often reluctant to turn to the Fund and subject themselves to conditionality, and, as noted above, the catalytic effect is only likely to be present in the case of countries which are already on the margin of creditworthiness. This is not the situation in low-income countries. One argument for offering subsidies on Fund lending is to provide an additional incentive for countries to turn to the Fund and import a Fund input in the design of macroeconomic policy. If the programmes of policies turn out to be successful, the reputation of IMF conditionality will improve and the catalytic impact of IMF conditionality will strengthen. In these circumstances the need for Fund lending will be reduced in the long run.

However, all this hinges on the appropriateness and effectiveness of Fund-supported programmes. Where IMF conditionality is not effective in encouraging balance of payments adjustment, the demand for balance of payments financing will remain high, the catalytic effect will weaken and the availability of private credit will diminish; the demand for finance from the Fund will therefore increase. Certainly for countries on the margin of creditworthiness Fund lending linked to effective conditionality may be necessary in the short run in order to reduce the need for it in the long run; Fund lending will induce economic adjustment and restore access to private capital markets.

Where adjustment programmes can achieve their objectives quite rapidly, the involvement of the Fund as a lending institution within particular countries will only need to be short-term. Hence we have the original vision of the Fund as a provider of temporary finance to deal with short-term balance of payments problems. However, where the size and nature of adjustment difficulties are more substantial, and where the catalytic effect is weaker, the Fund may be expected to be drawn into a longer-term lending role which unavoidably begins to overlap with that of development financing.

Central to the entire debate over the Fund as a lending institution is its effectiveness as an adjustment one. Within any one country, the more effective is IMF conditionality the less will be the long-run need for Fund lending. Assuming a constant flow of balance of payments shocks throughout the world, the aggregate level of Fund lending should not be expected to change, although the identity of the countries drawing on the Fund at any one time will change. In essence what one has here is the conventional trade-off between the speed of adjustment and the demand for international liquidity (see Clark, 1970, for one of the clearest statements of this trade-off). But superimposed on it is the additional element that where adjustment becomes slower and more difficult, the demand for loans from the IMF will become higher, not only because the overall demand for liquidity will rise but also because the supply of private loans will fall.

## General assessment

We have now answered the questions raised at the beginning of this section, even though on occasions the answer has been to explain why it is impossible to be precise. What has emerged is the following. If it is accepted that private markets fail, then there is a lending role for an international financial institution to play. Where the nature of the balance of payments problems facing countries is short-run, this lending function may be performed by the Fund within its existing Articles of Agreement. Where, however, balance of payments problems are of a longer-term and more structurally related type, the lending function of the Fund becomes less clear-cut since such problems may also be seen as developmental ones which require longer-term development finance; provision of this type of finance has conventionally been seen as lying beyond the Fund's legitimate role. The options are clearly either to channel such lending through a different IFI, or to legitimise this lending activity as part of the Fund's accepted role.

How large Fund lending should be clearly depends upon how the Fund's role is defined and upon how effective Fund conditionality is. However, once a role is defined it is possible to quantify, at least approximately, the resources that will be needed to carry it through. Such an approach has the attraction of focusing attention on the basic factors that determine the specific answer.

It is also difficult to reach a 'positive' answer to the question of whether the Fund's lending role should be different in different countries. The short answer is simply that it is different. Broadly speaking, there appears to be a graduation. As countries become richer they appear to draw less from the Fund (although again this is something that is tested empirically in the following chapter). At one extreme the developed countries do not borrow at all from the Fund. At the other extreme low-income countries seem often to be left with no option other than to draw from the Fund. And in between some countries at some times borrow from the Fund. As they become more fully developed, however, and make the transition from developing to developed status they may be expected to cease borrowing. Although the basic purpose of Fund lending is the same in all the countries that draw from it, the means by which this objective is realised may be expected to differ sharply, not least in terms of the duration of Fund involvement.

A logical case may be made for low conditionality Fund finance in circumstances where there is confidence that the government will design an appropriate adjustment strategy. On the other hand, to the extent that there is broad consensus about what this strategy should be, high conditionality should perhaps not be regarded as a significant cost, except perhaps in a political sense. The costs of high conditionality will clearly depend crucially on the nature and effectiveness of the conditions. Moreover, if it is accepted that it is the conditionality attached to Fund loans that makes them unique and which is needed to generate a catalytic effect, then the case for low conditionality finance weakens. It is only where the potential catalytic effect is unimportant that this argument against low conditionality will disappear. The difficulty is that this state of affairs will be most pronounced in low-income countries. Such countries may well require technical assistance with the design of macroeconomic policy, and in this respect Fund conditionality may have a significant role to play. But low-income countries are also likely to encounter severe supply-side problems, and this again creates difficulties for the Fund both in the design of adjustment policy and in terms of a longer-term lending commitment.

The political economy of conditionality is also likely to be very important. Unless it is simply the announcement of an agreement with the Fund that generates an impact by altering expectations and behaviour, the success of conditionality will depend on the extent to which programmes are implemented. Indeed, even where it is the announcement that is currently important, this will not remain the case since problems of time inconsistency will arise. The incentive to renege, itself created by the announcement effect, will eventually undermine the situation. But if conditionality will only work where it is implemented, the issue becomes one of the optimum level of conditionality in terms of maximising implementation. Even where low or no conditionality may be inappropriate it may be unwise to aim for conditionality that is perceived as being so strict that it disinclines countries from implementing it, or even from turning to the Fund for financial support in the first place. There is a real problem of conditionality optimisation. The political economy of conditionality suggests that it should certainly be no stricter than it absolutely needs to be.

The discussion of Fund lending conducted so far raises two further questions to which we now turn. The first is whether lending is more appropriately carried out through the General Resources Account, through various special facilities, or through the Special Drawing Rights Account. The second is how the Fund should itself be financed, if indeed it is to perform a lending function.

## SDRs AND OTHER FORMS OF FUND LENDING

SDRs have had a somewhat chequered history. In the mid-1970s the Fund envisaged that they might become the principal reserve asset in the international monetary system, and there were proposals for establishing substitution accounts to facilitate the transition away from the use of currencies as international reserves. At the same time, a good deal of attention was paid to making the SDR a more attractive asset in terms of its capital value, the interest it carried, and its usefulness and liquidity. It was against this background that considerable debate took place over the appropriate way of distributing SDRs, with developing countries favouring the establishment of a 'link' between SDR allocation and the provision of foreign aid.[37]

However, with the changes that occurred after the collapse of the Bretton Woods system in 1973, the problem of reserve adequacy, which had led to the introduction of SDRs in the first place, became systemically less important. Reserve adequacy was apparently no longer relevant in a world of flexible exchange rates and balance of payments financing by private capital markets. After 1981 no further SDR allocations were made and the world moved to a multiple currency system.[38]

A reversal of these systemic trends – the movement towards greater exchange rate management and the failure of private markets to provide a satisfactory means of balance of payments financing – is, however, creating new interest in SDRs. World economic recession is seen by some, not least the Fund's Managing Director, as reflecting a shortage of international liquidity and as justifying an additional allocation of SDRs. Indeed, throughout the 1980s some commentators have sought to defend SDR allocation not only in broad systemic terms but also in terms of helping to deal with specific problems such as developing country debt (Williamson, 1984) and the global environment (Bird, 1992).

Without replicating the full reviews of the SDR, the 'link', substitution accounts and the targeted use of SDRs that exist elsewhere, there are certain aspects of the SDR that are directly relevant to our discussion of IMF lending. First, drawings under the SDR account differ significantly from those under the General Resources Account and other Fund accounts. Strategically they are essentially unconditional and are not subject to any strict repayment schedule.[39] Second, as originally envisaged they were not intended to be a means of facilitating permanent real resource transfers, but were distributed on the basis of IMF quotas which were seen as standing as a proxy for the long-run demand to hold international reserves. Third, the potential for resource transfers under the SDR facility has been reduced by raising the interest rate on the net use of SDRs to a market-related level, but this has failed to eliminate their benefits to recipients. In many cases developing countries cannot borrow at market rates. Their lack of creditworthiness means that they face an availability constraint which the receipt and use of SDRs would help them overcome.[40] Broadly speaking the less creditworthy a country is, the more important SDRs will be. Considerations of political economy clearly reveal why, in a world where creditworthiness appears to be strongly and positively correlated with the level of development, little interest tends to have been shown in the SDR.

Where certain countries are identified as holding inadequate international reserves, the allocation of additional SDRs to them represents one way in which reserves may be raised to adequate levels. But beyond this simple statement lurk all the problems discussed earlier in this chapter.

If SDRs are only of use to countries that are not creditworthy, should the Fund be making resources available to them? Should such countries not instead be seeking to raise their creditworthiness by pursuing policies of economic adjustment? But then again will the receipt of SDRs help or hinder them in this endeavour? Where SDRs are spent and not held by recipients should the Fund become a source of quasi-permanent financial assistance? And in any case can one distinguish between balance of payments and development finance, and what is the appropriate timeframe to adopt?

Basically where the Fund is viewed as an agency for providing strictly temporary balance of payments finance, and where the unique attraction of Fund lending is the conditionality attached to it, the SDR facility will be seen as an inappropriate means of providing financial support. To those who see little distinction between balance of payments and development finance, who favour a longer-term perspective on the balance of payments, who see it as legitimate for the Fund to have a quasi-permanent financing function in low-income countries, and who favour low over high conditionality, the SDR – particularly where subsidised for some users – will represent an attractive means of Fund financing.

Compromises are in principle quite possible. SDR allocations could be made subject to strict reconstitution provisions; they could be allocated on a discretionary and not a universal basis and could be tied to conditionality.

However, a danger with the gradual and evolutionary reform of the SDR, and in particular a danger with channelling SDRs through the Fund's existing lending windows, is that the SDR will lose its basic simplicity of operation. A feature of reform has been the proliferation of facilities under which countries may borrow. Starting from the general principle that all circumstances were covered by conventional stand-by drawings, the situation has changed radically with more and more facilities being introduced to deal with specific difficulties. In the early 1960s the change seemed relatively undramatic and was represented by the introduction of the Compensatory Financing Facility, but this was followed by the Buffer Stock Financing Facility, the Extended Fund Facility, the Trust Fund, the Oil Facility, the Structural Adjustment Facility, the Enhanced Structural Adjustment Facility, the Compensatory and Contingency Financing Facility, not to mention various subsidy accounts.

What starts off as a sensible course of action to deal specifically with particular difficulties can be carried too far. The argument that there has been an excessive proliferation of lending facilities would be strengthened by empirical evidence that the facilities are not working well and by analysis which suggests that there are deficiencies in their design. Evidence to be presented in Chapter 3 raises serious questions about specific facilities. The ESAF and the remodelled CCFF have been little used since their introduction. The BSFF has never been greatly used during its life. The EFF was decreasingly used towards the end of the 1980s and seems to have had a particularly poor record of performance. The CCFF has lost the low conditionality features that used to be associated with the CFF before 1983, and yet has not operationally captured a uniqueness in terms of its contingency components.

In many instances researchers have found that, while the facilities may be differentiated in terms of the rubric that describes them, it is much more difficult to distinguish between them in the way in which they are used. If this is true, two options present themselves. The first is for the Fund to establish a clearer operational identity for each of its separate lending facilities. The second is to rationalise, to decide which distinctions are operationally significant and design a more limited number of facilities around these distinctions. Legitimate distinctions could, for example, be drawn between the appropriate level of conditionality and the appropriate type of conditions where high conditionality is deemed appropriate. But even these distinctions, which according to some interpretations are not at present significant in terms of the Fund's lending operations, could be accommodated within a less complex structure of lending facilities.

## FINANCING THE FUND: QUOTA-BASED SUBSCRIPTIONS AND THE ALTERNATIVES

If the Fund were to play only an adjustment role it would be a relatively inexpensive organisation to run, but if it is to make loans then it clearly requires the resources to lend. There are various ways in which these resources may be raised. The options are: subscriptions from member countries; borrowing from member countries; borrowing from private capital markets; the use of SDRs; and the sale of gold.

### Subscriptions from members

At present almost 70 per cent of the Fund's resources comes from members' subscriptions, which are determined by their quotas. Additional resources come from borrowing from members. Other Fund lending, such as that conducted through the Trust Fund, has been financed by sales of gold. Up to now the Fund has not borrowed directly from private capital markets.

A detailed analysis of the quota system has been made elsewhere by the author (Bird, 1987). A fundamental problem with IMF quotas is that they are used for many different purposes; they help determine voting rights, the size of ordinary drawing rights and access to special facilities, the size of SDR allocations to individual members, and the size of subscriptions to the Fund. Such multi-purpose quotas would not be a problem if the purposes were (perfectly) positively correlated, but in fact they may be in direct conflict. Raising quotas not only increases the supply of Fund resources but is also likely to increase the demand for them as well. Countries in the strongest position to provide the resources necessary to run the Fund are, almost by definition, unlikely to be the ones in greatest need of its financial support. Similarly, as noted earlier, the criteria for evaluating a country's need for support from the General Resources Account may not be exactly the same as those relevant for assessing its need for SDRs, since GRA resources are largely conditional liquidity whereas SDRs are essentially owned reserves.

Moreover, the apparently objective process by which quotas were initially set was in fact largely spurious, and the process through which they are increased is ill-defined and increasingly unsatisfactory in a world where global economic variables change rapidly. There is some empirical support for the claim that the outcome of quota reviews depends heavily on political factors and bargaining strength within the Executive Board of the Fund, since variations in the size of the quota increases cannot be explained simply in terms of key international economic variables (Bird, 1987). A lack of flexibility in altering quotas means either that the Fund's contribution to dealing with global economic problems will be constrained, or that other methods for increasing resources will have to be found. When faced with shortages of resources, the Fund, although subject to certain guidelines, has frequently resorted to direct borrowing from specific members as a means of increasing its capacity to lend. This has taken place under the General Arrangements to Borrow (GAB), the supplementary financing facility, bilateral borrowing arrangements with Saudi Arabia and Japan, and the policy on enlarged access.

In certain respects borrowing is a significant step away from the rigidity imposed by quotas. While still seeking to equate the aggregate demand for and supply of Fund resources, borrowing serves to alter the distribution of demand and supply. As a result some countries have become eligible to draw more resources from the Fund while subscribing no more resources to it; at the same time other countries have lent more resources to the Fund without an increase in their access to resources from it. In this way the Fund has acted as an intermediary in the process of international financial recycling, and the connection between quotas and both the demand for and supply of Fund resources has been relaxed. Some commentators have criticised borrowing from members as changing the basic nature of the Fund as a credit union, and altering the distribution of members' rights and obligations (Kenen, 1985).

There are further arguments to suggest that the mechanisms used by the Fund for escaping from the rigidity of the quota system are unsatisfactory and that greater reliance on suitably reformed quotas would be preferable. On the *demand* side these arguments relate both to the *cost* of those purchases from the Fund which are financed by borrowing, and to their *nature*. As regards cost, the charges on resources drawn under the Oil Facility, the SFF or the enlarged access policy (EAP) are higher than those applying to purchases under the Fund's other facilities, though Subsidy Accounts have been used in the case of the Oil Facility and the SFF in an attempt to assist the Fund's poorer members meet the higher charges.

With regard to their nature, because the SFF and the EAP have been used to 'top up' credit tranche stand-bys and EFF loans, they have meant that a higher proportion of Fund lending has been of a high conditionality type than would have been the case had quota increases been used to generate an equivalent amount of additional resources. This, of course, need not constitute a problem if Fund conditionality is perceived as being appropriate to those countries borrowing from the Fund, but this is not universally the case. The implications of the means of financing the Fund for the balance between high and low conditionality resources should not therefore be neglected.

On the *supply* side the main problem is the uncertainty associated with *ad hoc* measures. Can economically strong countries, and those in substantial balance of payments surplus – the identity of which may well change and become more dispersed – be relied upon always to provide the resources necessary for the Fund to maintain its programme of loans? While, given the institutional constraints and the immediacy of problems, borrowing may have been the only way of dealing with the Fund's own liquidity problems in the late 1970s and 1980s, it does not represent the best long-term solution for financing the Fund.

Yet similarly, and under the existing system of negotiated infrequent reviews, can quota increases which permit the Fund to play a central role in international economic affairs be guaranteed? If not, considerable costs are likely to ensue. If the resources are not forthcoming, there will be global costs in terms of higher levels of unemployment and lower levels of output and trade as countries are forced to adjust to lower levels of financing. Even if extra resources are eventually made available, the *uncertainty* regarding the outcome of negotiated quota reviews will clearly raise a question-mark over the role that the Fund can play and this may have a destabilising effect.

One way of ensuring that the Fund has sufficient resources is to index quotas against indicators of the need for Fund resources, bearing in mind that there are arguments for basing the distribution of potential access to Fund finance on a different set of criteria from that used for determining the distribution of subscriptions. By introducing a more automatic and direct link between the need for Fund resources and their supply, many of the costs associated with infrequent reviews could be avoided.

Moreover, the effects of global inflation on the real value of quotas could be more easily neutralised. As things stand, quotas are expressed in nominal SDRs; inflation therefore means that from the very moment a new set of quotas is ratified, their real value begins to fall. Clearly the quantitative relevance of this depends on the rate of inflation; acute in the mid-1970s, the problem became less significant in the 1980s, even though other reasons for questioning the adequacy of quotas associated with different aspects of global economic performance became more marked. Index linking is, of course, a fairly conventional means of trying to reduce some of the costs of inflation. But if the global quota is to be linked to broader indicators of global economic performance, what should these indicators be?

One option is to use the value of world trade. The ratio of quotas to world trade has fallen significantly from 14.2 in 1950 to 11.5 in 1960, 8.2 in 1971 and 3.8 in 1981. In 1990 it was 3.5. Although the precise value of the ratio depends on whether the year chosen is just before or just after quotas have been raised, the downward trend is well established. However, although the quota:trade ratio is convenient to calculate, it is a poor proxy for the adequacy of Fund resources, since the demand for these is related to the incidence and size of payments deficits, which may not be perfectly positively correlated with the *level* of world trade. Indeed, the use of the quota:trade ratio may be criticised for basically the same reasons as those assembled against using the reserves:imports ratio as a measure of reserve adequacy, the most telling being that the simple use of ratios fails to provide a rigorous explanation of what constitutes the optimum value for the particular ratio chosen, or indeed what factors most significantly influence this value. This leads on to the question of whether it is possible to gain any insight into assessing the adequacy of Fund resources and therefore quotas from other approaches to the adequacy of international reserves.

In the case of Fund quotas there are in fact additional problems on top of those normally associated with judging reserve adequacy. In part, these relate to the cost of producing extra Fund resources, which require countries to swap foreign exchange and SDRs for Reserve Positions in the Fund and to subscribe more of their own currency to the Fund at a potential future real resource cost. But they also relate to the benefits of Fund resources, since these are affected by the fact that a large proportion of Fund resources are conditional; the benefits therefore depend on the appropriateness of the conditions attached to Fund loans. Moreover, the benefits of extra Fund resources also depend on the global advantages of expanded Fund activity. While it is possible to talk about these in general terms, it is difficult to convert them into a satisfactory and objective specific value. In relation to this latter point only a small part of Fund resources may be counted as 'reserves'; the rest represents credit.

A conclusion from the above discussion is that Fund resources have distinctive features as compared with other forms of reserve assets and international liquidity. Their conditionality, along with the fact that Fund operations may involve positive externalities, needs to be borne in mind when assessing the adequacy of Fund resources.

Faced with the problems of quantification, analysis of the adequacy of international reserves has also drawn on a qualitative or symptomatic approach. The basic idea behind this is as follows: a shortage or excess of reserves will exert an impact on certain key economic variables either directly, through, for instance, affecting the domestic money supply, or more indirectly by encouraging the pursuit of particular policies. By observing the *policies* that are pursued and the performance of certain key *economic variables*, one may reach tentative conclusions about reserve adequacy. This approach has also been subjected to considerable criticism (Bird, 1985), and again its deficiencies are multiplied if adopted as a means of assessing the adequacy of Fund resources. The question is not simply one of whether there are enough reserves or whether there is enough international liquidity but also one of whether there is the right balance *between* the various reserve assets and components of international liquidity; does the system need more Fund-based resources *relative* to the other types of international finance? Although it is possible, in principle, to derive a symptomatic guide to answering this question – for example, by examining the appropriateness of macroeconomic policies in individual countries or by looking at the implications of private bank financing – it is difficult to convert such generalities into specific values for Fund quotas.

While this brief review suggests no easy answers to the problem of determining the adequacy of the global quota, it does help to identify some of the factors that should be taken into account. It has, in other words, identified some of the arguments in the implicit demand function for Fund resources. As shown by the author elsewhere (Bird, 1987), a greater degree of objectivity could be introduced into calculating the value of resources that the Fund requires and translating this into appropriate subscriptions and quotas.

## Borrowing from private lenders

Just as the World Bank finances some of its activities by private borrowing, could the IMF not avail itself of this source of finance? Would commercial banks, for example, lend to the Fund? Although this again would alter some of the basic features of the Fund, direct borrowing from commercial banks does, in principle, offer a way of recycling resources from surplus countries to deficit ones; there would simply be two intermediaries involved in the transaction. The incentive for banks to lend to the Fund would have to arise from the rate of return offered and their own assessment of the relative risk involved, which would, of course, be influenced by the way in which the Fund planned to use the extra resources. If the banks assessed the risks associated with lending to the Fund as being less than those involved in direct lending to the eventual recipients of Fund credit, they might be prepared to accept a relatively low rate of return.

However, it might encourage them to pull out of direct payments financing. Indeed, the most common criticism of Fund borrowing from private markets is that it would crowd out other borrowers, including developing countries, and possibly the World

Bank. In this context it needs to be recognised that debt crises have in a sense done this anyway, and the counter-argument can be made that in aggregate terms Fund borrowing might well crowd in additional financial flows; even so, the concern has to be taken seriously.

Another problem with direct borrowing relates to its cost. Would the Fund simply on-lend at the price and for the time period on which it borrowed, or would it attempt to transform the maturity and reduce the interest rate? Where the Fund charged a lower rate on its lending than it paid on its borrowing, an additional financing problem would clearly arise. However, if it did not do so, direct borrowing from private markets would be of more limited benefit to many developing countries.

Finally, too heavy a concentration on private borrowing as a means of financing the Fund would expose its activities to the vagaries of the capital market. Moreover, resort to borrowing from the markets might persuade some governments to reduce their own financial support for the Fund, with the result that it might become even more difficult to get agreement on quota increases.

In conclusion, and bearing in mind the public-good nature of many of the Fund's activities, it seems more appropriate that these should be supported by government subscriptions.[41]

## Using SDRs and gold

A third alternative, which is in certain ways more attractive than either of the above, is for the Fund to finance its activities by the creation of SDRs which would then be transferred to the General Account.[42] The attractions are associated with increasing the significance of the SDR and allowing the Fund to use its own asset more fully. Furthermore, quotas, in their role of determining individual country subscriptions, could be dispensed with. Such a move would, of course, mean that some SDRs would be issued to Fund members in the form of conditional and repayable credit – under the auspices of the GRA. It would also rely heavily on the SDR becoming a more useful asset. Finally, there is the implicit presupposition that it would be easier to get governments to agree to extra SDR allocations to the General Account than to quota increases. Although countries would not be asked to give up financial resources directly, they would still be a potential real resource cost. More to the point, if governments oppose any expansion in the Fund's activities, they may be expected to resist this by whatever means it is to be financed. Such resistance may also be encountered against a plan to finance the Fund's lending activities through further sales of gold.[43]

## CONCLUDING REMARKS

IMF lending raises a myriad of difficult analytical issues. It touches on questions of efficiency and equity, and involves both positive and normative economics as well as considerations of international political economy and the economics of bureaucracy and public choice. Precise and uncontentious answers should not be expected; they certainly cannot be given. Indeed, the more precise the answers, the more contentious they also generally become.

If one believes that international equity is unimportant and that markets are always efficient, or at least more efficient than governments, there is reasonable justification for arguing that neither the IMF nor any international agency should be lending at all. We have argued, however, that private international capital markets have shown themselves to be inefficient on the basis of a number of criteria. Moreover, the international community should not ignore those countries that are ignored by private markets. International distributional issues are important. In any case uncreditworthy countries still have balance of payments problems. In these circumstances there is a role for lending by international financial institutions. But should it be the IMF?

This chapter shows how the original blueprint for Fund lending, designed as it was for the purpose of short-term balance of payments financing, encounters difficulties where payments correction becomes a longer-term process relying heavily on supply-side change. This means that Fund lending, conducted as conventionally envisaged, will often be inappropriate. The increasing overlap between balance of payments and development finance has made the purpose of Fund lending increasingly unclear and has been reflected in the proliferation of Fund lending facilities. Operationally, however, the success of this institutional response must be seriously challenged. Rhetoric has differed from practice.

The distinguishing features of Fund loans, in principle, are their cost and their conditionality. Indeed, Fund loans may need to be concessionary in order to encourage borrowers to accept the conditionality that is attached to them, and to be appropriate to the debt-servicing capacity of poorer ones. Where Fund-supported programmes are successful a rather strong case for Fund lending may be made. Where they are unsuccessful it becomes more difficult to support Fund lending except as a form of aid, and then the question is whether loans from the Fund are the best way of providing aid. Certainly what emerges is that the Fund as a lending institution cannot be divorced from the Fund as an adjustment institution. The more efficient it becomes in this latter role, the less will be the need for it in its former role, although, at the same time, the stronger will be the justification for using the availability of loans from the IMF as a way of encouraging countries to pursue Fund-supported adjustment programmes.

Traditionalists will argue that the Fund should not be lending where there is little or no chance of reasonably rapid balance of payments correction. Indeed, the strongest case against Fund lending does seem to be that the nature of international financing gaps has changed in a way that reduces the relevance of conventional forms of Fund lending. Two policy responses suggest themselves. The first calls for significant adaptation on the part of the Fund to modify the ethos of its lending policies. The second calls for existing and future financing needs to be met in ways that do not require the Fund to alter its basic orientation. These options will be discussed further in Chapter 4. In the following chapter, however, we move on to an empirical investigation of IMF lending, which draws on the analytical issues introduced in this one.

# APPENDIX
## Modelling IMF lending and conditionality

Public choice theory assumes that the Fund seeks to maximise its power and influence.

These ultimate objectives are served by maximising the following utility function:

$$U^F = f(C,L)$$

where $C$ = conditionality, $L$ = lending, and

$$\frac{dU^F}{dC} > 0,$$

and

$$\frac{dU^F}{dL} > 0$$

Potential borrowers' utility is a function of the same arguments,

$$U^B = f(C,L)$$

but for them:

$$\frac{dU^B}{dC} < 0$$

and

$$\frac{dU^B}{dL} > 0$$

The preference maps of the Fund and potential borrowers may be illustrated by Figure 2.1.

In negotiation both the Fund and the borrower are seeking to achieve their highest possible utility function. At a given level of conditionality $C^1$ they would both prefer to have greater use of Fund credit ($Z > Y > X$). However, where increased lending is associated with increased conditionality, an increase in the Fund's utility may coincide with a decline in that of the borrower, compare

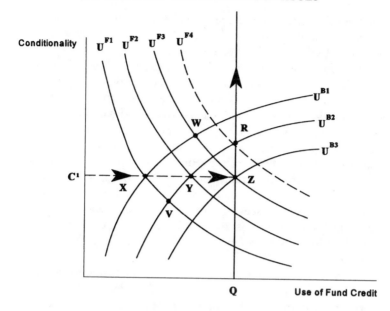

*Figure 2.1* Preference maps of the IMF and potential borrowers

$V$ on $U^{F1}$ $U^{B2}$ with $W$ on $U^{F3}$ $U^{B1}$; although Pareto efficient gains are still possible in those circumstances, compare $Z$ with $V$.

Lending will, however, be constrained by a country's access limits; there will be a quota constraint as shown by $Q$ in Figure 2.1. In these circumstances the combination of conditionality and lending shown by $Z$ is Pareto efficient; raising the level of conditionality would increase the Fund's utility but would decrease that of the borrower, and vice versa. Through increasing the utility of Fund credit to borrowers, a decline in the level of conditionality will increase their demand for it. However, were the Fund to seek stricter conditionality in an attempt to raise its own utility to $U^{F4}$, this would reduce the utility for borrowers to $U^{B2}$ and the demand for Fund credit would tend to fall, thereby eroding the marginal utility gained by the Fund.

The model presented above is not without its attractions as a schematic device. The relationships it spells out can be assessed against the empirical evidence, and the model can be made more sophisticated. It can, for example, be modified to allow for the conditionality rises, and if the costs of non-compliance are perceived by the borrower as insignificant or even falling as conditionality rises, then the borrower's utility functions may bend upwards rather than downwards. From the Fund's side it is more difficult to see how non-compliance fits in; for if the borrower complies with the Fund's conditions and its balance of payments improves, the power and influence of the Fund will, in one sense, be only temporary, although the 'status' of the Fund as an effective financial institution may increase. To the extent that non-compliance reduces effectiveness, it will prolong the Fund's involvement but will diminish its status. Besides, where are the power and influence of the Fund if borrowers ignore its conditions? If the Fund is seeking to maximise compliance, but at the same time compliance is a negative function of the level of conditionality, this represents another trade-off for the Fund to deal with.

The model does, however, tellingly rest on restrictive assumptions, in particular with regard to the Fund's objective function. An alternative hypothesis is that the Fund seeks to *minimise* the level of conditionality it perceives as being necessary to induce a targeted improvement in the balance of payments of borrowers. If the Fund believes that a relaxation in conditionality will have positive incentive effects, its utility will then become a negative function of conditionality, although conditionality would, in these circumstances, be more appropriately seen as an intermediate as opposed to a final good. Moreover, any attempt to present the Fund as a unified actor misrepresents the diversity of view that exists within the organisation.

# 3

# IMF LENDING
## The empirical evidence

### INTRODUCTION

Leading on from the previous chapter, we now examine various empirical aspects of lending by the IMF. Rather than attempting to provide evidence relating to every aspect, we try instead to identify the major trends that may be discerned. We also try to draw on the empirical evidence where this helps to resolve some of the analytical questions raised in Chapter 2.

Concentrating initially on the General Resources Account (GRA) and other special facilities for low-income countries, the chapter investigates the size and pattern of Fund lending, disaggregating this both over time and across countries. Having provided an overall picture of Fund lending, we then use the available evidence to see whether it is possible to come up with a statistically secure explanation of what determines the use of Fund resources, incorporating both global and country-specific factors.

Next we look at the relative importance of Fund lending both in terms of the size of balance of payments deficits that member countries have been facing and in terms of financial flows from other sources. We then move on to examine the concessionality and conditionality of Fund lending and explore the extent to which these may have influenced drawings on the Fund and may be used to offer increased assistance to low-income countries. A concluding section then examines the SDR Account and investigates the extent to which developing countries have been net users of SDRs.

### THE SIZE AND PATTERN OF FUND LENDING

As noted in Chapter 1, the IMF has in some ways 'changed partners' in the period since the mid-1970s. At the beginning of

97

the 1970s industrial countries were still drawing substantial amounts from the Fund and often accounted for the clear majority of Fund lending in value terms. Even as late as 1974–7 industrial countries were important clients of the Fund, with the United Kingdom and Italy making relatively large drawings. During the 1980s, however, Fund lending was exclusively to developing countries, with the economies of Central Europe and the former Soviet Union becoming important users of Fund finance at the beginning of the 1990s. Table 1.1 succinctly shows the evolving pattern of net credit from the Fund during 1982–92. At the beginning of the 1980s the Fund was most heavily involved in lending to African and Asian developing countries, although only the minority of loans to Africa went to countries that were sub-Saharan. Moreover, lending was not in the main going to countries producing primary products. In fact, three times as much credit was provided to developing countries where manufactures constituted the predominant export.

The debt crisis in 1983 had a radical impact on the pattern of Fund lending. The highly indebted countries of Latin America now became major users of Fund finance, while loans to Africa actually fell. Indeed, in summary terms, the entire increase in Fund credit between 1982 and 1983 was accounted for by the 15 most heavily indebted developing countries, with most of these to be found in Latin America. Other aspects of Table 1.1 confirm the impact of the debt crisis on Fund lending; credit to 'market borrowers' increased by almost 300 per cent, and credit to countries with recent debt-servicing difficulties rose from $4 bn in 1982 to $7.9 bn in 1983, while credit to those without such difficulties rose merely from $2.9 bn to $3.1 bn. Clearly there seems to be adequate proof that the debt crisis opened a new chapter in relations between the IMF and developing countries.

However, 1983 turned out to be a peak year for Fund lending. Net credit to developing countries fell from $11 bn in 1983 to only $4.7 bn in 1984 and thereafter to almost zero in 1985. This general pattern existed across all categories of developing countries, although the near-zero figure in 1985 concealed small positive net credits to countries in the Western Hemisphere, in essence the heavily indebted countries, and net repayments to the Fund by Asian and European developing countries. Certainly, then, the relatively high net flows of finance from the IMF to developing countries in 1983 were not sustained. Indeed, for the rest of the period from 1986 to 1992, with the exception of 1991, developing countries paid more back to the Fund than they received in new loans; this was the situation for both debt-distressed and other developing countries alike.

In 1991, while net credit to countries in transition in Central and Eastern Europe was $3.5 bn, net credit to all other developing countries was only $1.1 bn. Within this latter total, net credit to African developing countries was a modest $0.2 bn, and net credit to developing countries in the Western Hemisphere and heavily indebted countries was again negative, following a year when positive flows seemed to have been re-established. By 1992 net credit to countries in transition had fallen to $1.7 bn and net credit in Africa had turned negative.

Further information on IMF lending is shown in Table 3.1 which provides data on outstanding commitments rather than net credit. The table again shows the collapse in Fund lending in the mid-1980s from the peak reached in 1983. However, it also shows how the trough in lending running from 1985 to 1989 appears itself to have been a temporary interlude, with outstanding commitments rising sharply in 1989–90 and continuing to rise more smoothly up until the end of 1992. Yet even then, at around SDR 19 bn, outstanding commitments were some way below the SDR 25 bn level reached in 1983. At the same time, whereas there were 39 arrangements in effect in 1983, this had increased to 53 in 1992. The clear conclusion is that the amount of finance associated with each programme on average fell – a trend associated with the introduction of SAFs and ESAFs. Moreover, not shown in the table is the fact that, while the *stock* of outstanding commitments increased over the entire 1989–91 period, the *flow* of *new* commitments fell by about 50 per cent in 1990–1 by comparison with the previous year. Stock figures can therefore give a misleading impression of current flows.

Information on the classification of IMF lending by individual facilities is also contained in Table 3.1. Significant trends may again be identified. At the beginning of the 1980s Fund lending was fairly equally divided between stand-bys and Extended Fund Facility (EFF) loans. But most of the expansion in lending up to 1983 occurred under the EFF; over 70 per cent of new commitments in 1981–2 were made under this facility. The situation then changed rapidly and radically and EFF loans accounted for only 2 per cent of new loans in 1983–4. With some exceptions in 1985–6,

Table 3.1 IMF arrangements in effect in financial years ended 30 April, 1953–92

| Financial year | Number of arrangements as of 30 April | | | | | Amount committed as of 30 April (in millions of SDRs) | | | | |
|---|---|---|---|---|---|---|---|---|---|---|
| | Stand-by | EFF | SAF[a] | ESAF | Total | Stand-by | EFF | SAF[a] | ESAF[b] | Total |
| 1953 | 2 | | | | 2 | 55.00 | | | | 55.00 |
| 1954 | 3 | | | | 3 | 112.50 | | | | 112.50 |
| 1955 | 3 | | | | 3 | 112.50 | | | | 112.50 |
| 1956 | 3 | | | | 3 | 97.50 | | | | 97.50 |
| 1957 | 9 | | | | 9 | 1,194.78 | | | | 1,194.78 |
| 1958 | 9 | | | | 9 | 967.53 | | | | 967.53 |
| 1959 | 11 | | | | 11 | 1,013.13 | | | | 1,013.13 |
| 1960 | 12 | | | | 12 | 351.38 | | | | 351.38 |
| 1961 | 12 | | | | 12 | 416.13 | | | | 416.13 |
| 1962 | 21 | | | | 21 | 2,128.63 | | | | 2,128.63 |
| 1963 | 17 | | | | 17 | 1,520.00 | | | | 1,520.00 |
| 1964 | 19 | | | | 19 | 2,159.85 | | | | 2,159.85 |
| 1965 | 23 | | | | 23 | 2,154.35 | | | | 2,154.35 |
| 1966 | 24 | | | | 24 | 575.35 | | | | 575.35 |
| 1967 | 25 | | | | 25 | 591.15 | | | | 591.15 |
| 1968 | 31 | | | | 31 | 2,227.36 | | | | 2,227.36 |
| 1969 | 25 | | | | 25 | 538.15 | | | | 538.15 |
| 1970 | 23 | | | | 23 | 2,381.28 | | | | 2,381.28 |
| 1971 | 18 | | | | 18 | 501.70 | | | | 501.70 |
| 1972 | 13 | | | | 13 | 313.75 | | | | 313.75 |
| 1973 | 12 | | | | 12 | 281.85 | | | | 281.85 |
| 1974 | 15 | | | | 15 | 1,394.00 | | | | 1,394.00 |

| Year | | | | | | | | | | |
|---|---|---|---|---|---|---|---|---|---|---|
| 1975 | 12 | | | | 12 | 337.25 | | | | 337.25 |
| 1976 | 17 | 2 | | | 19 | 1,158.96 | 284.20 | | | 1,443.16 |
| 1977 | 17 | 3 | | | 20 | 4,672.92 | 802.20 | | | 5,475.12 |
| 1978 | 19 | 3 | | | 22 | 5,075.09 | 802.20 | | | 5,877.29 |
| 1979 | 15 | 5 | | | 20 | 1,032.85 | 1,610.50 | | | 2,643.35 |
| 1980 | 22 | 7 | | | 29 | 2,340.34 | 1,462.85 | | | 3,803.19 |
| 1981 | 22 | 15 | | | 37 | 5,331.03 | 5,464.10 | | | 10,795.13 |
| 1982 | 23 | 12 | | | 35 | 6,296.21 | 9,910.10 | | | 16,206.31 |
| 1983 | 30 | 9 | | | 39 | 9,464.48 | 15,561.00 | | | 25,025.48 |
| 1984 | 30 | 5 | | | 35 | 5,448.16 | 13,121.25 | | | 18,569.41 |
| 1985 | 27 | 3 | | | 30 | 3,925.33 | 7,750.00 | | | 11,675.33 |
| 1986 | 24 | 2 | | | 26 | 4,075.73 | 831.00 | | | 4,906.73 |
| 1987 | 23 | 1 | 10 | | 34 | 4,313.10 | 750.00 | | 327.45 | 5,390.55 |
| 1988 | 18 | 2 | 25 | | 45 | 2,187.23 | 995.40 | | 1,357.38 | 4,540.01 |
| 1989 | 14 | 2 | 23 | 7 | 46 | 3,054.05 | 1,032.30 | 954.97 | 1,566.25 | 6,607.57 |
| 1990 | 19 | 4 | 17 | 11 | 51 | 3,597.02 | 7,834.40 | 1,370.20 | 1,109.64 | 13,911.26 |
| 1991 | 14 | 5 | 12 | 14 | 45 | 2,702.58 | 9,596.70 | 1,812.95 | 539.42 | 14,651.65 |
| 1992 | 22 | 7 | 8 | 16 | 53 | 4,832.63 | 12,158.85 | 2,110.73 | 101.15 | 19,203.36 |

*Source:* IMF, *Annual Report*, 1992.

*Notes:* [a] Includes arrangements where the three-year commitment period has expired but the third annual arrangement remains in effect (three cases in 1991 and two cases in 1992). The committed amounts exclude these cases.
[b] Includes amounts previously committed under SAF arrangements that were replaced by ESAF arrangements.

the EFF continued to play a very modest role up until the end of the 1980s. During 1988–9, for example, 65 per cent of new lending by the Fund was in the form of stand-bys, 21 per cent in the form of ESAF loans, 10 per cent in SAF loans and a mere 5 per cent in the form of EFF loans. However, the proposition that the EFF window was being allowed to close was challenged by events at the end of the 1980s. In 1989–90 almost 70 per cent of new commitments were entered into under the auspices of the EFF and, by the end of 1992, 63 per cent of outstanding Fund commitments were under seven EFF arrangements.

To what extent does the changing relative importance of individual facilities reflect the country composition of Fund lending? Is it that low-income countries tend to draw under the SAF and ESAF arrangements, while other developing countries draw under stand-bys and the EFF? As at the end of 1991, 24 per cent of IMF commitments were to low-income countries, 55 per cent to other developing countries, and 21 per cent to former Comecon countries. Examination of a full record of those countries that borrowed from the IMF during 1985–90 shows that, although extended arrangements have existed with some low-income countries such as Ghana and Malawi, EFF lending has been dominated by loans to Brazil, Chile, Mexico and Venezuela. Loans to these countries have frequently absorbed a very large proportion of total IMF lending. In 1990, for example, each of the EFF loans to Mexico and Venezuela individually exceeded the total stand-bys agreed with 19 other countries. Again, each of these loans on its own easily exceeded the total of SAFs and ESAFs agreed with 22 countries. The four arrangements negotiated with Eastern European countries during 1985–90 (two of which were with Hungary) all took the form of stand-by arrangements.

While the changing geographical composition of Fund use certainly appears to influence the extent to which individual facilities are used, it is also the case that the Fund's own attitudes to individual facilities have changed over time and this has probably also affected their usage. For some time after the early 1980s the Fund's management showed considerable scepticism regarding the EFF, believing that its record was relatively poor and that the longer-term involvement of the Fund could be better achieved via a series of shorter-term programmes. This no doubt had some bearing on the extent to which the facility was used. A 'tightening up' of the EFF towards the end of the 1980s reawakened the Fund's enthusiasm for it.

At the beginning of the 1990s the picture was one of a few relatively large EFF loans and a larger number of relatively small loans under other facilities. In 1990 the average value of each EFF in effect was SDR 1.96 bn (1,958.5 m.), whereas the average value of stand-bys was SDR 189.3 m., of SAFs SDR 65.3 m. and of ESAFs SDR 124.7 m. By 1992, although there were many fewer EFF arrangements than stand-bys, SAFs, or ESAFs, the amount committed under EFFs was significantly higher.

As already explained, the range of IMF facilities so far described all involve a high degree of conditionality. Reforms to the Compensatory Financing Facility (CFF) have, however, also increased the degree of conditionality attached to what was historically a relatively low conditionality lending window. While expected to co-operate with the Fund in finding solutions to their payments difficulties, countries which experienced export shortfalls which were largely beyond their control had reasonably automatic and rapid access to finance through the CFF. In the mid-1970s the CFF represented a relatively important source of Fund finance for developing countries. Even in 1982 drawings under the CFF accounted for some 30 per cent of total drawings on the Fund. However, an internal review of the CFF in 1983 resulted in stricter guidelines for judging 'co-operation' and reduced access limits to the facility. Most independent analyses suggest that these changes meant that the CFF was no longer a low conditionality facility, and 1984 saw a significant decline in drawings under it. With the exception of 1987, drawings under the CFF (and the CCFF as it became) never rose above $1 bn, and by the end of 1990 they were barely more than $1 m.[44] Net purchases were positive for 1980–3 but were negative during 1984–8. The geographical pattern of drawings across developing countries approximately matches the pattern of drawings in general, with African countries being the principal users of the CFF in 1981 and 1982, but with countries in the Western Hemisphere taking over this role thereafter, making net positive drawings under the facility up until 1985.

Certainly it would seem that the evidence on the use of the CFF, and yet more so the CCFF, is consistent with the view that increasing conditionality has made the facility less attractive to potential users. Aware of this, the Fund in 1993 revised the rules relating to the CCFF, although there remained no intention of introducing soft contingency lending.

The only remaining lending facility to examine is the Buffer Stock Financing Facility (BSFF), but its use during the 1980s has been so low as hardly to warrant mention. During the entire decade only $500 m. were drawn under the BSFF, and there were zero net drawings in 8 of the 11 years in the period 1980–90.

The relatively low drawings under both the CCFF and the BSFF raises the question of whether they are facilities worth having. The principal distinguishing feature of the CFF as a low conditionality facility has been lost, while the BSFF is essentially redundant. In such circumstances it might seem sensible either to rethink the role of the CCFF and endeavour more clearly to distinguish it from other IMF facilities or, if this is not done, simply to amalgamate it with other facilities.[45]

Before moving on to offer some interpretation of the trends observed above, it is relevant to explore two other aspects of Fund lending. Some analysts have interpreted references in the Fund's Articles of Agreement to the 'temporary' and 'revolving' character of Fund lending as implying that the Fund is a credit union (Kenen, 1985). But is it the case that Fund involvement is temporary? Or, on the other hand, does the Fund tend to have prolonged relationships with the countries to which it lends? Evidence quoted in Chapter 1 from Goreux (1989) showed how low-income countries in particular find it difficult to disengage themselves from the Fund, having once turned to it for financial assistance. Some low-income countries have had outstanding credit from the Fund for almost 30 consecutive years. Indeed, Goreux discovered that 21 countries had credit outstanding from the Fund for in excess of 14 years. Such continuous Fund involvement clearly challenges one of the basic tenets of its operations. But is prolonged financial support by the Fund limited to low-income countries? Figure 3.1 provides an answer to this question by showing the number and time duration of IMF stand-by arrangements running from 1985 to 1990. During this period 47 countries negotiated stand-bys. Of these, 25 had only one stand-by, but the remaining 22 had at least two and sometimes three. In some of the cases where only one stand-by was arranged, drawings on the Fund were also made under other facilities. Mexico, for example, while having only one stand-by covering the end of 1986, 1987 and the beginning of 1988, also had extended arrangements which covered

1984–5 and 1989–92. Mexico was therefore involved with the Fund throughout almost the entire period. Indeed, of the 25 countries that had only one stand-by, five drew resources from the IMF under its other facilities as well. Of the 10 countries that had three stand-by arrangements between 1985 and 1990, six were low-income countries, but four were not.

This evidence seems to confirm that the image of the Fund coming into a country, offering swift financial support, helping to turn the balance of payments around, and then getting out, is purely and simply wrong. Moreover, a relationship that spans a number of years is not purely a feature of the Fund's dealings with low-income countries. It applies to better-off developing countries as well, and seems likely to apply to the economies of Central and Eastern Europe.[46] The quasi-continuous involvement of some countries with the Fund implies either that there are few costs of failing to comply with conditions in terms of eligibility for future loans, and in this sense the commitment to adjustment could be adversely affected, or that IMF conditionality where observed, is not very effective. Certainly a more well-articulated system of incentives with rewards for good behaviour and penalties for bad behaviour might be worthy of examination. Current conditionality would then in part reflect a country's previous record. Moreover, reforms to raise the effectiveness of conditionality would reduce individual countries' long-term reliance on the Fund.

One final aspect of Fund lending which deserves attention is the growing problem of arrears. Having never previously been a worry for the IMF, during the 1980s and particularly towards the end of the 1980s arrears became a serious issue (see Table 3.2). By 1992, ten countries owed a total of approaching $5 bn, equivalent to almost an eighth of the Fund's total outstanding credits. For many of these countries the arrears covered a period of more than three years. Two aspects of the problem are noteworthy. First, the IMF was now exposed to the delays in repayment that had been experienced by other creditors and this could do little other than influence its own lending policies. The Fund had to try and maintain its monetary reputation, and arrears did not help. The claim that conditionality was necessary to ensure repayment and to guarantee the revolving nature of Fund lending which had sounded somewhat hollow now had to be taken more seriously. Second, the arrears almost exclusively related to low-income countries or the poorer developing countries with this further suggesting that their

*Figure 3.1* The persistence of Fund involvement in developing countries, 1984–92[a]

| Stand-bys | 1984 | 1985 | 1986 | 1987 | 1988 | 1989 | 1990 | 1991 | 1992 |
|---|---|---|---|---|---|---|---|---|---|
| Argentina | 12 | — | 3 | 7 — | 9 | 11 — | — | 3 | |
| Belize | 12 | — | 3 | | | | | | |
| Central African Republic | | 9 | — | 3  6 — | 5 | | | | |
| Congo | | 8 | — | — | 8 | | 8 | — | 5 |
| Costa Rica | 3 | 4 | | 10 — | — | 3  5 — | 5 | | |
| Côte d'Ivoire | 6 — | 6  6 | — | — | 6 | 11 — | — | 4 | |
| Dominican Republic | | 4 — 4 | | | | | | | |
| Ecuador | 3 — | 3 | | | 1 — | 2  9 — | — | 2 | |
| Egypt | | | | 5 — | 11 | | | | |
| El Salvador | | | | | | | 8 — | 8 | |
| Equatorial Guinea | | 6 — 6 | | | | | | | |
| Gabon | | | 12 — | — | 12 | 9 — | — | 3 | |
| Gambia | | | 9 — 10 | | | | | | |
| Ghana | | 8 — | 12 | 10 — | 10 | | | | |
| Guatemala | | | | | 10 — | — | 2 | | |
| Guinea | | | | 2 — | 3  7 — 8 | | | | |
| Guyana | | | | | | | 7 — | 7 | |
| Haiti | | | | | | 9 — | 12 | | |
| Honduras | | | | 3 — | 5 | | 7 — | 8 | |
| Hungary | | | | | 5 — 5 | | | | |
| Jamaica | | 7 — | — | 5 | 9 — | 11 — | 5 — | 5 | |
| Jordan | | | | | | 7 — | — | 1 | |
| Kenya | 2 — | 2 | | | 2 — | 7 | | | |
| Korea | | 7 — | 3 | | | | | | |
| Liberia | 12 | — | 6 | | | | | | |
| Madagascar | | 4 — | 4  9 — | — | — | 2 — | 7 | | |
| Malawi | | | | | 3 — | 5 | | | |
| Mauritania | | 4 — | 4 — | 4 | | | | | |
| Mexico | | | 11 — | — | 4 | | | | |
| Morocco | | 9 | 12 | 2 | 3  8 — | 12 | 7 — | 8 | |
| Nepal | | 12 | — | 4 | | | | | |
| Niger | 12 — | 12 — | 12 — | 12 | | | | | |
| Pakistan | | | | | 12 — | — | 11 | | |
| Panama | | 7 — | — | 3 | | | | | |
| Papua New Guinea | | | | | | | 4 — | 6 | |
| Philippines | 12 | — | 6  10 | — | — | 4 | | | |
| Poland | | | | | | | 2 — | 3 | |
| Senegal | | 1 — | 7 | 10 — | 10 | | | | |
| Sierra Leone | | 11 — 11 | | | | | | | |
| Somalia | 2 — | 2 | 6 — | — | 2 | | | | |
| Thailand | 6 | — | 3 | | | | | | |
| Toga | 5 — | 5  6 | — | — | 4 — | 4 | | | |
| Trinidad and Tobago | | | | | | 1 — | 2  4 — | 3 | |
| Tunisia | | 11 — | — | 5 | | | | | |
| Yugoslavia | 5 — | 5 | | | 6 — | 6 | 3 — | 9 | |
| Zaire | | 5 | — | — | 3 — 5 | 6 — | 6 | | |
| Zambia | 7 — | 4 | | | | | | | |

*Figure 3.1* (continued)

| Extended arrangement | 1984 | 1985 | 1986 | 1987 | 1988 | 1989 | 1990 | 1991 | 1992 |
|---|---|---|---|---|---|---|---|---|---|
| Brazil | — | — | 2 | | | | | | |
| Chile | | 8 — | | | — 8 — | 8 | | | |
| Ghana | | | | 11 — | | | — 11 | | |
| Malawi | — | — | 9 | | | | | | |
| Mexico | — | 12 | | | | | | | |
| Philippines | | | | | | | 5 — | | — 5 |
| Tunisia | | | | | | 7 — | | — 7 | |
| Venezuela | | | | | | | 6 — | | — 6 |

| Structural adjustment | 1984 | 1985 | 1986 | 1987 | 1988 | 1989 | 1990 | 1991 | 1992 |
|---|---|---|---|---|---|---|---|---|---|
| Bangladesh | | | | 2 — | | | — 2 | | |
| Bolivia | | | 12 — | | | — 12 | | | |
| Burundi | | | 7 — | | | — 6 | | | |
| Central African Republic | | | | 6 — | | | — 5 | | |
| Chad | | | | 10 — | | | — 10 | | |
| Dominica | | | 11 — | | | — 11 | | | |
| Gambia | | | 9 — | | | — 9 | | | |
| Ghana | | | | | | | | | |
| Guinea | | | | 7 — | | | — 7 | | |
| Guinea Bissau | | | | 10 — | | | — 10 | | |
| Haiti | | | 12 — | | | — 12 | | | |
| Kenya | | | | | 2 — | | | — 2 | |
| Lesotho | | | | | 6 — | | | — 6 | |
| Madagascar | | | | 8 — | | | — 8 | | |
| Mali | | | | | 8 — | | | — 8 | |
| Mauritania | | | 9 — | | | — 9 | | | |
| Mozambique | | | | 6 — | | | — 6 | | |
| Niger | | | 11 — | | | — 11 | | | |

*Note:* [a] The numbers refer to the months of the year.

economic problems are even more deep-seated and fundamental than those facing other developing countries. Again, the question arises as to how best the IMF may assist such countries.

The analysis of Fund lending in this section reveals a number of things. First, it shows that large swings in Fund lending occur, but that to a considerable degree these are associated with the fact that the amount of Fund lending at any one time is dominated by a relatively small number of relatively large arrangements. In terms of the overall number of arrangements, an upward trend since the early 1970s is more firmly established, even though there was a sharp acceleration in the number of loans at the beginning of the 1980s followed by a deceleration in the mid-1980s. The Fund is apparently fairly persistently *widening* its involvement.

Table 3.2 Arrears to the Fund: members with obligations overdue by six months or more (SDR m.) year ended 30 April

|  | 1986 | 1987 | 1988 | 1989 | 1990 | 1991 | 1992 | 1993 |
|---|---|---|---|---|---|---|---|---|
| Amount overdue | 489 | 1,186.3 | 1,945.2 | 2,801.5 | 3,251.1 | 3,377.7 | 3,496.0 | 30,064 |
| Number of members | 8 | 8 | 9 | 11 | 11 | 9 | 10 | 12 |
| of which | | | | | | | | |
| Gen. Dept | 418.9 | 1,088.4 | 1,787.7 | 2,594.2 | 3,018.6 | 3,171.7 | 3,274.1 | 27,683 |
| No. of members | 8 | 8 | 9 | 11 | 11 | 9 | 10 | 12 |
| SDR Dept | 12.2 | 15.6 | 25.1 | 35.0 | 44.7 | 27.3 | 37.5 | 49.5 |
| No. of members | 5 | 4 | 6 | 6 | 9 | 6 | 7 | 9 |
| Trust Fund | 57.9 | 82.3 | 132.4 | 172.3 | 187.8 | 178.7 | 184.3 | 188.3 |
| No. of members | 6 | 6 | 7 | 7 | 9 | 6 | 6 | 6 |
| No. of ineligible members | 4 | 5 | 7 | 8 | 10 | 8 | 8 | 7 |

Source: IMF Annual Report, 1993.

Second, the decline in Fund lending both in terms of the number of arrangements and the amount committed over the period 1983–6 coincided with a time when many developing countries were encountering severe economic problems in the wake of the debt crisis, and this calls into question whether the Fund has performed its 'public-good' role of providing required finance which the private markets are unwilling or unable to provide. Third, the large changes in the use of some facilities, most pronounced in the case of the EFF, and the low usage of others, such as the CCFF and BSFF, suggest that there is no clearly identifiable strategy lying behind the current range of IMF facilities, and that there is scope for significant modification and rationalisation.

## EXPLAINING FUND LENDING: BROAD FACTORS

The pattern and trends of IMF lending may, in principle, be explained by a combination of demand-side and supply-side factors. On the demand side, countries will not need to borrow from the Fund, and, in any case, will be ineligible to do so, unless their balance of payments is in deficit. Even with a balance of payments deficit, countries may opt to avoid the Fund. They may prefer to pursue their own programme of adjustment independently of the Fund, or they may decide to finance their deficit by running down reserves or by borrowing from other sources. It may therefore be expected that the demand for loans from the Fund will tend to rise as balance of payments deficits become larger, as the nature of conditionality becomes more acceptable to potential borrowers, and as the availability of alternative sources of payments financing falls or its cost rises. Globally, it will be the distribution of deficits and not just their size which is important.

Undoubtedly an important factor in explaining growing balance of payments disequilibria world-wide during the 1980s has been the large and increasing US current account deficit. But one would not expect this to have any discernible or *direct* impact on IMF lending, since the US remains internationally creditworthy. For those countries whose creditworthiness has evaporated, as happened to the highly indebted developing countries after 1983, there may be little alternative but to turn to the Fund. It is useful to recall that in Chapter 2 a simple typology of countries, which emphasised creditworthiness, was suggested to explain the pattern of Fund lending across countries and over time. For

Table 3.3 Evolution of maximum access to Fund resources (as % of quota)

| Facility or policy | May 1947 | Feb 1963 | Sep 1966 | June 1969 | | Sep 1974 | | Apr 1975 | | Dec 1975 | | Jan 1976 | | Apr 1978 | |
|---|---|---|---|---|---|---|---|---|---|---|---|---|---|---|---|
| | | | | A | B | A | B | A | B | A | B | A | B | A | B |
| Credit tranche | 100 | 100 | 100 | 100 | 100 | 100 | 165 | 100 | 165 | 100 | 165 | 145[a] | 176.25 | 100[b] | 165 |
| EFF | – | – | – | – | – | 140 | – | 140 | – | 140 | – | 140 | – | 140 | – |
| Compensatory Financing Facility (exports) | – | 25 | 50 | 50 | – | 50 | 75 | 50 | 75 | – | 75 | – | 75 | – | 75 |
| Compensatory Financing Facility (cereal) | – | – | – | – | 75 | – | 75 | – | 75 | – | – | – | – | – | – |
| BSFF | – | – | – | 50 | – | 50 | – | 50 | – | 50 | – | 50 | – | 50 | – |
| Oil facility (1974) | – | – | – | – | – | – | 75 | – | 75 | – | – | – | – | – | – |
| Oil facility (1975) | – | – | – | – | – | – | – | – | 125 | – | 125 | – | – | – | – |

Table 3.3 (continued)

| Facility or policy | Feb 1979 | | Dec 1980 | | May 1981 | | Jan 1984 | | Dec 1984 | | Dec 1986 | | Dec 1989 | |
|---|---|---|---|---|---|---|---|---|---|---|---|---|---|---|
| | A | B | A | B | A | B | A | B | A | B | A | B | A | B |
| Credit tranche | 100[b] | | 100 | | 100 | | 100 | | 100 | | 100 | | 100 | |
| Extended Fund Facility | 140 | 165 | 140 | 165 | 140 | 165 | 140 | 165 | 140 | 165 | 140 | 165 | 140 | 165 |
| SFF/ | 102.5[c] | 305[d] | 500[c] | 600 | 500[c] | 600 | 308[c]–400 | 408–500 | 308[c]–350 | 408–450 | 300[c]–340 | 400–440 | 300[c]–340 | 400–440 |
| enlarged access | 140[e] | | 460[e] | | 460[e] | | 268[e]–360 | | 268[e]–310 | | 260[e]–300 | | 260[e]–300 | |
| Compensatory Financing Facility (exports)[f] | 100 | | 100 | 100 | 100 | 125 | 83 | 105 | 83 | 105 | 83 | 105 | 83 | 122 |
| Compensatory Financing Facility (cereal)[f] | | | – | | 100 | | 83 | | 83 | | 83 | | 83 | |
| Contingency financing[f] | – | | – | | – | | – | | – | | – | | 65 | |
| BSFF | – | 50 | – | 50 | – | 50 | – | 45 | – | 45 | – | 45 | 45 | 45 |
| SAF[g] | – | | – | | – | | – | | – | | – | 47 | | 70 |
| ESAF | – | | – | | – | | – | | – | | – | | | 250 |

Source: IMF, Treasurer's Department.

Notes: A indicates maximum by type of access; B indicates maximum combined access.

[a] In January 1976, the Fund extended temporarily the size of each credit tranche by 45%. From January 1976 the Fund applied its credit tranche policies on the basis of the increased credit tranches, each equal to 36.25% of quota.

[b] Following the Second Amendment in April 1978, the size of each of the credit tranches reverted to 25% of quota.

[c] Stand-by arrangement.

[d] This limit could be exceeded in exceptional circumstances.

[e] Extended Fund Facility.

[f] In August 1988 the compensatory financing facility was replaced by the CCFF.

[g] Access increased by 63.5% in July 1987 and to 70% in March 1989.

industrial countries creditworthiness is well established and at the very least will take some time to erode. The evidence suggests that such countries do not borrow from the Fund. For low-income countries persistently poor creditworthiness determines an equally persistent demand for IMF resources. We do indeed empirically observe the relatively large number of Fund arrangements with such countries, not only under stand-bys but also increasingly under the structural adjustment windows. For middle-income countries, such as those in Latin America, the picture is less clear-cut because their creditworthiness is uncertain and varies over time. While they were creditworthy in the 1970s and early 1980s they avoided the Fund, but as their creditworthiness declined so their demand for Fund resources increased.

On the supply side of the Fund lending equation there are, first of all, the access limits which constrain the ability of countries to draw under various facilities (these are shown in Table 3.3); second, the Fund's willingness to lend under different conditions; and third, the charges that the Fund makes on the use of its credit. Certainly there have been occasions when an increase in IMF lending has been directly associated with the liberalisation of a particular facility or an increase in quotas. Moreover, the implementation of stricter conditionality by the Fund will mean that more programmes are rejected, or that the negotiation of the programme breaks down, or that countries, mindful of the conditionality they are likely to encounter, opt not to turn to the Fund in the first instance.

In terms of access limits, the Ninth General Review of Quotas implemented in 1993 resulted in a 50 per cent increase in quotas. At the same time the Fund terminated its Enlarged Access Policy, under which it had supplemented its quota resources with borrowed funds. It also reduced its access limits 'broadly to maintain potential access to IMF financing'. The new limits are shown in Table 3.4, although the Fund maintains that 'in exceptional circumstances' it may approve stand-bys or extended arrangements for excess amounts. The limits are reviewed annually.

A further more complex supply-side factor has been the Fund's policy towards commercial lending and latterly to arrears. Immediately after the onset of debt crisis the Fund's policy of concerted lending attempted to link the provision of its own resources with additional commercial assistance. As this lapsed the Fund insisted that recipients of its loans remained current with their commercial

*Table 3.4* Access limits under IMF arrangements, 1993 (in % of quota)

|  | Under old quotas | Under new quotas |
| --- | --- | --- |
| Access under credit tranches and the Extended Fund Facility |  |  |
| Annual | 90–110 | 68 |
| Cumulative | 400–440 | 300 |
| SAF[a] | 70 | 50 |
| 1st Year | 20 | 15 |
| 2nd Year | 30 | 20 |
| 3rd Year | 30 | 15 |
| ESAF[a] |  |  |
| Maximum | 250 | 190 |
| Exceptional | 350 | 255 |
| CCFF | 122 | 95 |
| Sublimits: |  |  |
| Compensatory[b] | 40 | 30 |
| Contingency | 40 | 30 |
| Cereal[b] | 17 | 15 |
| Optional tranche | 25 | 20 |
| BSFF | 45 | 35 |
| Augmentation for debt/ debt-service reduction | 40 | 30 |

*Source: IMF Press Release*, No. 92/81.

*Notes:* [a] Access over a three-year period.
[b] If the balance of payments position, apart from the effects of the export shortfall (cereal import costs), is satisfactory, the current limit is 83% of quota, and the new limit is 65% of quota.

debt obligations; hence, in some quarters and as noted in Chapter 1, it was viewed as a debt collector for the banks. Such a policy may clearly have disinclined countries from borrowing from the Fund. It is noteworthy that the relaxation of this policy towards the end of the 1980s coincided with an increase in Fund lending.

On a quantitatively more muted scale a similar impact might be associated with the Fund's evolving policy on arrears. Here the Fund has adopted an 'intensified collaborative strategy', which, according to one well informed observer, may be seen as 'the ultimate in disguised rescheduling' (Kafka, 1991). Under this strategy a member country which agrees to follow a programme monitored by the Fund will receive help in obtaining bank credits and other loans to clear its arrears. Thereafter, and with its eligibility to draw on the Fund restored, it will enter into a regular financial arrangement with the Fund to provide enough resources to repay bridging credits. Part of the strategy is that the country can earn 'rights' to larger drawings from the Fund than would normally be available. While limited to countries that had protracted arrears at the end of 1989, such policy changes can clearly, at a disaggregated level, have a significant impact on Fund lending, although by early 1993 only Peru had successfully completed the process. The Fund's insistence that its credits cannot be officially rescheduled, let alone forgiven, clearly constrains the response it can make to low-income countries in arrears.

This discussion illustrates why explaining the overall pattern of Fund lending is always likely to involve a complex combination of demand-side and supply-side factors. But can any broad suggestions be offered to help interpret the swings in lending during the 1980s and early 1990s that were observed in the previous section? At the outset it needs to be recognised yet again that the concept of Fund lending is itself complex. Are we talking about the stock of outstanding credit, the flow of new loans, or net purchases, i.e. new loans minus repayments? Moreover, should we be looking at the value of IMF lending or the number of loans? Some figures will illustrate the problem. In 1977 there were 20 high conditionality arrangements in effect which had a value in committed resources of about SDR 5.5 bn. By 1980 the number of arrangements had increased to 29 but the value of committed resources had fallen to under SDR 4.0 bn. By 1983, with 39 arrangements in place, commitments had increased dramatically to just over SDR 25 bn. Five years later, in 1988, there were 45 arrangements, but the value of commitments had plummeted to only SDR 4.5 bn. These observations show that it is misleading and unwise to examine Fund lending at too highly aggregated a level. A rise in the value of loans may go hand in hand with a fall in their number as larger economies turn to the Fund and smaller countries turn away. In order to explain the overall picture we therefore need to explain the behaviour of each set of countries.

In terms of the period 1980–92, three phases of Fund lending seem to warrant explanation. The first is the dramatic increase in lending that occurred in the early 1980s. The second is the rapid and pronounced decline and subsequent trough in lending during the mid-1980s. And the third is the partial resurgence of lending at the beginning of the 1990s.

In 1981 almost all IMF lending was to low-income countries. Both Asian and more particularly African countries experienced a marked decline in their current account balance of payments in 1981; in the case of Africa the deficit rose from just over $1 bn ($1,028 m.) in 1980 to nearly $22 bn ($21,894 m.) in 1981. Adjustment could not quickly be achieved; holdings of international reserves were low; and private financing was meagre. Net total bond lending to Africa was actually negative in 1981, and international bank lending was only $2.4 bn, representing just over 10 per cent of net resource flows. Low-income countries had nowhere to turn to help them with their payments needs, apart from the IMF.

While IMF lending to low-income countries remained fairly static until 1983, there was a big increase in lending to Latin American countries. There can be little doubt that this was an exclusively demand-led phenomenon; indeed, if anything, supply-side factors, and in particular an increased emphasis on conditionality, disfavoured increased lending. While normally this would have reinforced many Latin American countries' reluctance to borrow from the Fund, the environment was not normal. In 1983 the current account balance of payments deficit in developing countries in the Western Hemisphere had actually fallen quite dramatically from the levels of 1981 and 1982. Thus in 1983 it stood at about $8 bn ($8,166 m.) in comparison with almost $42 bn ($41,641 m.) in 1982. But it was their deteriorating debt situation, and more particularly the rapid decline in private financial flows associated with their falling creditworthiness, which forced them into the Fund. Given the economic size of the indebted Latin American countries, there was a dramatic impact on the value of IMF lending, though not, one may note, on the number of loans.

To some extent a decline in new lending and net credits might have been expected to follow this sudden surge as countries began to repay the loans, but, as shown in Table 3.1, it was not only new loans but also outstanding commitments that fell. A combination of offsetting factors seems to have been at work. While private financial flows continued to decline for most developing countries, which would have led them to look for alternative sources of finance, two other factors pulled in the opposite direction. The first was the improvement in their current account balance of payments which strengthened significantly in 1984 and 1985 by comparison with 1983, and although it worsened in 1986 it was actually in surplus in 1987. This clearly had an impact on the demand for finance from the Fund.

Second, although the Fund talked more about growth and about longer-term structural change, there are also indications that in some ways conditionality became stricter as the 1980s progressed. There is, for example, evidence that the proportion of programmes which involved devaluation rose from just under 60 per cent in 1973–80 to 82 per cent in 1981–3, and to practically 100 per cent after 1983 (Polak, 1991). Devaluation has been notoriously unpopular with developing countries, although perhaps less so over recent years. There is a reasonable presumption that immediately following the debt crisis in late 1983 the IMF was anxious to avoid a general international banking and financial collapse. Although it was keen to emphasise adjustment as the appropriate response to the crisis, the strategic importance of the highly indebted countries encouraged it to relax conditionality in some cases. However, the evidence in general is that conditionality was rather strict during the 1980s. The conventional caricature of a Fund-supported programme which places heavy reliance upon controlling domestic credit creation and encouraging exchange rate devaluation became still more accurate (Edwards, 1989) and, at the same time, programme elements began to relate to a wider range of policy variables. Evidence suggests that the average number of performance criteria per programme rose from less than 6 during 1968–77 to 9½ between 1984 and 1987 (Edwards, 1989). While from one point of view this could be interpreted as the Fund aiming for a richer and more appropriate mix of policy, it could also be seen by potential borrowers as a more detailed and deeper, and therefore more unwelcome, intrusion by the Fund into the design of domestic policy – something to be avoided if at all possible.

The only potential puzzle here is the counter-claim often made by the Fund itself, and to some extent acknowledged by area experts, that developing countries in Latin America began to accept the wisdom of IMF orthodoxy. Why, if they accepted the thrust of IMF advice, did they not borrow more heavily from the Fund? Surely the perceived adjustment cost of Fund finance had been reduced? The answer could be twofold. First, the timing of the conversion to IMF orthodoxy could have occurred at the end of the 1980s and the beginning of the 1990s when there is an indication of increased lending activity by the Fund. Second, even where IMF advice seems more acceptable, countries still regard the Fund as a lender of last resort. The idea that countries rush to the Fund just as soon as they can in order to get access to its resources does not stand up to the empirical evidence.

What certainly does seem to have been the case during the mid to late 1980s is that the Fund had unused lending capacity, even though quotas had not been increased since 1983.[47] Lending arrangements (excluding the CCFF) as a percentage of quota fell from 214 per cent in 1980 to 115 per cent in 1982. Although they rose to 165 per cent in 1983, they declined dramatically thereafter, falling to a mere 56 per cent in 1987 and 57 per cent in 1988.[48] Fund lending was not supply-constrained.

Turning now to the increase in lending at the beginning of the 1990s, it is difficult to see how this was linked to any marked change in conditionality, although, as noted above, perhaps the Fund's major clients began to believe that IMF-type stabilisation programmes were unavoidable and that exchange rate realignment was necessary. Perhaps, at the same time, pronouncements by the Fund, relating to structural adjustment and medium-term balance of payments viability, the protection of vulnerable groups and social safety nets, created a softer image for the institution, irrespective of the reality. Probably more important than either of these influences was the large deterioration in the current account balance of payments of developing countries as a whole, their combined deficit increasing from $19.0 bn in 1989 to $104 bn in 1991. The weakening balance of payments position was a feature of all regional sub-groups, affecting primary product producers as well as producers of manufactured goods, and straddling developing countries with and without debt-servicing difficulties. On top of this, increasing claims from Central and Eastern European economies began to make a significant impact on the level and pattern of IMF lending. Again demand-side factors appear to predominate.

It is also interesting to note that, while the stock of arrangements and the amount committed under them rose during 1989–92, the size of arrangements approved during each financial year varied significantly, the numbers again being dominated by a few large arrangements. In 1989 one EFF was arranged with a value of SDR 207.3 m., whereas in 1990 three EFFs committed more than $7 bn ($7,627.1 m.). Swings in IMF lending, therefore, can often be accounted for by whether individual large developing countries made drawings on the Fund. Again for this reason demand-side factors appear to lie at the heart of explaining the size and pattern of Fund lending, although in principle the Fund has the ability to exert some influence over demand via its conditionality and the charges it makes, issues to which we return later. Before doing so can we be more precise about the types of countries which borrow from the Fund?

## THE ECONOMIC CHARACTERISTICS OF USER COUNTRIES

Is it possible to define the economic characteristics of countries that do and do not draw on the Fund? Is it possible in some sense to quantify a demand function for loans from the Fund?

It might be expected that there will again be no straightforward answers to these questions – an expectation that turns out to be quite rational. There is a substantial and growing amount of evidence that use of Fund resources is a 'political' as well as an economic phenomenon. Countries with some form of economic or political 'clout' can, so it appears, often negotiate more preferable terms than those that are weak and lack influence (Stiles, 1990). Knowledge that there are differences in bargaining power, and that this counts in terms of the deal that is struck, may be expected to exert an influence over the demand for IMF credits. Moreover, if, as appears to be the case, an application for financial support from the Fund is not infrequently associated with natural disasters such as droughts, hurricanes, cyclones, earthquakes and floods, it will remain rather difficult to predict the pattern of Fund lending until our ability to predict such meteorological and geological phenomena improves. Perhaps we shall end up with a sunspot theory of Fund lending!

Attempts to model the demand for Fund loans have met with only limited success (these are more fully analysed and extended in the Appendix to this chapter). One early study (Bird and Orme, 1981) used regression analysis to see whether a statistical relationship could be established between drawings on the Fund and key country economic characteristics, including the balance of payments, the debt–service ratio, the rate of inflation, *per capita* income, the level of reserves, the value of imports, and access to private bank credits. The model fitted the data well for 1976, with the estimated coefficients being consistent with *a priori* reasoning for all the explanatory variables except external debt, for which the coefficient was in any case not significantly different from zero. Developing countries seemed to draw more from the Fund as their balance of payments deteriorated and their inflation rate accelerated. They seemed to draw less as they became richer. Private borrowing and borrowing from the Fund seemed to be complementary activities and, to this extent, some statistical support for a catalytic effect of IMF lending on private lending was discovered, although Bird and Orme also concluded that drawings from the Fund and the acceptance of the related conditionality were neither a necessary nor a sufficient condition for access to private loans. However, the model broke down for subsequent years and was largely unsuccessful when modified in an attempt to predict which developing countries would and would not draw from the Fund. Taking a sample of 27 developing countries which did not draw on the Fund, the model predicted zero drawings in only eight cases. Something seemed to be holding countries back. Generally the authors concluded that their model omitted important political, social and institutional factors.

A rerun of a similar econometric model for 1980–5 (reported more fully in the Appendix) produces rather similar results; the coefficients on the inflation, income, and balance of payments variables are all statistically significant and have the expected sign. The coefficient on commercial lending is also significant but only weakly positive, again casting some doubt on a strong catalytic effect. Overall, however, the equation fits the data rather badly and certainly forms a poor basis upon which to predict Fund drawings. Indeed, a scaled version of the model yields a coefficient of determination of 23 per cent; 77 per cent of the variation in borrowing from the Fund remains unexplained by the model.

Another study of the early 1980s reaches a broadly similar conclusion while using a somewhat different econometric methodology (Joyce, 1992). Here a logit analysis of 45 countries that did and did not negotiate programmes with the Fund is undertaken covering the period 1980–4. Estimation of the model suggests that countries that are relatively poor, are pursuing relatively expansionary domestic economic policies as proxied by the growth of domestic credit and the size of the government sector, have more acute payments problems, and possess depleted reserves, are more likely to turn to the Fund than are other countries. The mean values of the five significant macroeconomic variables are shown in Table 3.5. The inflation coefficient, while positive, is found to be insignificant; whether inflation results in drawings on the Fund would seem to depend upon whether it results in payments problems. Interestingly, while also statistically insignificant, private

*Table 3.5* Mean values of economic characteristics in countries borrowing and not borrowing from the IMF

| Variable | Programme countries | Non-programme countries |
|---|---|---|
| Domestic credit growth | 35.5% | 28.4% |
| Government spending/ GDP | 0.160 | 0.135 |
| Current account/ exports | −0.318 | −0.225 |
| Reserves/imports | 0.185 | 0.315 |
| Per capita GDP | $944 | $1,491 |

*Source:* Joyce (1992).

flows are found to be negatively related to Fund lending, thus perhaps hinting that IMF lending is a substitute for commercial lending and again calling into question the notion of a catalytic effect, at least during the early 1980s.

Also interesting is the fact that this study found IMF drawings to depend only insignificantly, but negatively, on the debt–service ratio. This confirmed the finding made earlier by Bird and Orme and by the re-estimation of a modified form of their model for 1980–5 (see Appendix). A high burden of debt does not seem necessarily to lead to drawings on the IMF. Even the suggestion that it is perceived creditworthiness rather than debt alone which influences drawings on the Fund finds scant statistical support, with Joyce discovering only weak and insignificant relationships between Fund lending and various measures of creditworthiness. Although the individual coefficients may tell us something about the demand for Fund resources, the demand equation constructed by Joyce also has low predictive qualities and, in this respect as well, is consistent with the one estimated by Bird and Orme (1981).

Support for the claim that econometric studies have yet to specify a satisfactory demand equation is also provided by attempts by Cornelius (1987a, 1987b) to estimate the demand for IMF credits by sub-Saharan and by other non-oil developing countries. In his study of sub-Saharan Africa, Cornelius estimates his model over two sub-periods, 1975–7, and 1981–3. Although over the 1975–7 period it performs satisfactorily, it throws up a number of slightly puzzling results. As in the Joyce study, inflation appears to have an insignificant effect on the demand for IMF credits, although the sign is negative. Of greater interest is the fact that for both the 1975–7 and 1981–3 periods the coefficient on the balance of payments is insignificant. Indeed, in the latter period the sign is negative. Also noteworthy is that Cornelius finds a negative coefficient on his private lending term in both sub-periods, thus again challenging the idea of a catalytic effect, at least as far as Africa is concerned. For the 1981–3 period he finds only one coefficient that is significantly different from zero, and the coefficient of determination suggests that key explanatory variables are missing. Attempts to provide these in terms of exchange rate policy proved difficult to model, but Bird and Orme had found that this did not appear to help to explain IMF drawings, and Cornelius is led to conclude, as they did, that IMF drawings are not a purely economic variable.

Further work by Cornelius (1987c) tests for the idea that variations in IMF conditionality have significant quantitative effects on the demand for IMF credits. Although Bird and Orme had suggested that many of the developing countries in their study which did not use Fund resources would have encountered only low conditionality had they drawn, and that there was therefore some other factor disinclining them from drawing, Cornelius finds that variations in the strictness of conditionality over time do exert a considerable impact on Fund lending. Having estimated an equation for the demand for Fund credits from a period when conditionality was fairly liberal, he calculates what level of drawings would have been expected in a year when conditionality was more strict. He finds that drawings are significantly lower than would have been expected on the basis of his estimated equation.

A final, as yet unpublished, study by Conway (1991) has discovered much better predictive ability, with the explanatory variables used in his model accounting for almost 70 per cent of the variations in country participation in IMF-backed programmes. Based on a sample of 73 developing countries, 53 of which participated in an IMF-backed programme, and covering a period running from 1976 to 1986, Conway attempts to explain participation in terms of a combination of past performance, proxied by economic growth and the balance of payments, and the contemporaneous external economic environment, proxied by the terms of trade and the external debt burden. His results suggest that all these factors made a significant contribution to explaining observed participation, although he claims that credit rationing and sluggish export growth made it more likely in the 1980s than beforehand. He finds participation negatively related to past economic growth and the recent strength of the balance of payments, and also to real interest rates, with this perhaps suggesting credit rationing at lower rates, and to the terms of trade, but positively related to the stock of external debt, in particular short-term debt. Using the share of output from the agricultural sector as a proxy for the level of development, he also discovers a negative relationship between participation in IMF-backed programmes and the level of economic development. It is also noteworthy that he identifies an inertial component to participation; those countries that have borrowed from the Fund in the past seem more likely to borrow from it in the future.[49] Of course, while such persistent participation occurs it becomes difficult to disentangle the effects of economic performance on participation from the effects of participation on performance.

Where does all this leave us in terms of explaining IMF lending? Clearly there are some inconsistencies between the various studies, although probably no more than might have been expected given the different time periods investigated, country groupings examined and econometric methodologies adopted. But, at the same time, some fairly general and durable conclusions emerge. First, drawings on the Fund are generally associated with prior adverse balance of payments performance, with important elements in this being both domestic economic mismanagement and deteriorating terms of trade. Second, the IMF is a relatively more important source of finance for relatively poor countries. Economic growth and development can wean a country away from the Fund. The Fund therefore needs seriously to consider the interface between development and the balance of payments. Third, the much vaunted catalytic effect finds rather patchy empirical support, at least in terms of the effect of IMF lending on lending by private financial markets. Although concerted lending may have linked IMF lending more directly to commercial lending in the period following the debt crisis, claims that the catalytic effect is important and has grown need to be treated carefully and examined further in the context of the impact of Fund lending on official aid flows.[50] If a catalytic effect exists it is clearly not a blanket effect covering all countries, all time periods, and all other sources of finance. It is more subtle than this and almost certainly varies across the variables listed above. Thus IMF lending may have little impact on commercial flows in Africa but may influence aid flows, whereas in key Latin American developing countries which are on the margin of creditworthiness the Fund's catalytic role could be important for commercial flows.[51] Of course, the motivations behind aid from donors' perspectives vary and although many donors may be expected to take IMF and World Bank involvement and the related conditionality into account, it is perhaps unlikely to have a clearly discernible impact on their own lending policies. A review of the lending policies of a large range of aid donors by Hewitt and Killick (1993) seems to confirm this. But for now there remains little econometric support for the catalytic effect.

Fourth, the empirical evidence suggests that, while there may be some aspects of Fund lending which imply elements of a moral hazard problem, the strong moral hazard argument against Fund lending is not supported. Thus the inertial aspect, whereby countries develop a long-term involvement with the Fund, may suggest that there is little penalty for failing to comply with programmes that are negotiated in terms of diminished access to future resources from the Fund, and this may reduce governments' commitment to see programmes through. On the other hand, there are many reasons why programmes may fail, and lack of commitment is only one. Evidence presented in Chapter 1 shows that low-income countries, in particular, are exposed to adverse exogenous factors which may blow programmes off course. Moreover, it should also be noted that there is evidence to suggest that the degree of compliance makes relatively little difference to the success of a programme (Killick, *et al.*, 1992).

Turning to the strong moral hazard argument, the evidence is of excess supply rather than excess demand. Countries do not rush to the Fund. The fact that countries borrowing from the Fund usually face extreme economic difficulties does not necessarily mean that they contrived to achieve such situations in order to gain access to Fund resources, but is more likely to mean that it is only in such extreme circumstances that they are reluctantly forced into the Fund. Where other options exist countries will normally seek to avoid borrowing from the IMF.

Fifth, the notion that the Fund seeks to influence the demand for its credits by changing its conditionality receives little support.

If countries turn to the Fund only as a last resort and when other options do not exist it may be expected that they would have to accept whatever conditions the Fund lays down, a tendency that would be encouraged by a belief that failure to achieve conditions will not impair future access. Although developing countries with strong bargaining positions because of their strategic importance have sometimes been able to negotiate relatively lenient programmes, the public choice notion of the Fund varying conditionality over time in the light of the strength of demand for its credits does not appear to be widely substantiated (see below).

A final conclusion from the econometric analysis of Fund lending is that there is much that we do not understand about the circumstances under which countries will and will not make use of Fund credit; the high residual in most econometric studies suggests that there are probably a range of other non-economic factors which still need to be delineated.[52] But can the Fund do more to encourage countries to use its resources and to manage the size and pattern of usage by adjusting its conditionality and what it charges for such use?

## LENDING BY THE FUND: CONCESSIONALITY AND CONDITIONALITY

It was noted in Chapter 2 that some critics of Fund lending have painted a picture of countries contriving to create the balance of payments crises that will enable them to gain access to relatively low-cost Fund finance; they have argued that there is a moral hazard associated with IMF lending. At the same time, none of the studies cited in the previous section have attempted to estimate the effect that variations in Fund charges have on Fund lending either through time or even across countries, often because it has been claimed that the grant element has not varied significantly over the period studied. However, it is an issue that warrants some further examination since it would be unwise to ignore the price variable in an evaluation of demand. Furthermore, it may be useful to widen the interpretation of 'price' to include the degree of conditionality.

Two things are clear from an examination of the data relating to Fund charges. First, charges have risen fairly persistently over the entire period since 1950. Second, and more significantly, the gap between the Fund's basic rate of charge and comparable private market rates has narrowed markedly since the first half of the 1980s, the period covered by many of the studies reported in the previous section. During 1981–5 the basic rate of charge on the Fund's ordinary resources was 6.35 per cent as against the 6-month Eurodollar rate of 12.64 per cent. Clearly over this time period the grant element on IMF credit was significant and there was a price incentive to borrow from the Fund as opposed to private markets. During 1986–90, however, while the Fund's rate rose to 6.57 per cent, the Eurodollar rate fell to 7.98 per cent. By late 1990 the differential had been almost completely eliminated with the Fund rate standing at 8.30 per cent and the Eurodollar rate at 8.33 per cent. The period since 1980 has therefore seen significant variations in the concessionary (or grant) element in Fund lending.

As can be seen from these data, the variations in the extent of effective subsidy on Fund resources by comparison with commercial loans have primarily resulted from relatively large swings in commercial rates. It was high commercial rates in the early 1980s (as well as in the late 1960s and mid-1970s) that gave IMF borrowing a high concessionary element at these times, while it was the fall in commercial rates in the second half of the 1980s that largely eradicated it. Moreover, it needs to be remembered that, with the movement towards structural adjustment lending in the form of the EFF, loans from the Fund might now be regarded as medium-term, and therefore it may be inappropriate to compare Fund rates with short-term commercial rates. By 1993 Fund rates were probably 2–3 percentage points below the minimum medium-term commercial rate at which developing countries might borrow.

Although it is difficult to calculate exactly what impact the increase in the relative cost of Fund borrowing had on the Fund's lending activity, the movements in the two series are superficially consistent; Fund lending did fall as the interest rate on IMF loans increased and as commercial rates declined, and the concessionary element therefore fell. There was a rapid increase in new arrangements with the Fund under stand-bys and the EFF over 1981–3 when the concessionary element reached its historical peak. Interestingly other peaks in Fund lending such as the mid-1970s also coincided with relatively high concessionary elements. Certainly by the end of the 1980s it would only have been the existence of credit rationing in private markets or a positive desire by countries to be exposed to Fund conditionality that would have enticed them to borrow from the IMF as opposed to the private international financial markets. Of course, credit rationing may be very relevant. For many developing countries creditworthiness remained low as therefore did their access to private credit at market rates, and for this reason high Fund rates and a low concessionary element may not have deterred borrowing. It should also be noted that the rapid increase in new lending by the Fund in 1990 took place at a time when the grant element was low. It is simply too glib to argue that because the fall in the grant element on IMF loans in the mid to late 1980s coincided with a decline in IMF lending the one *caused* the other. It is, for example, quite possible that falling world interest rates contributed to expanding world economic activity which alleviated the balance of payments problems that developing countries faced, reduced debt-service payments, and thereby reduced the *demand* for IMF credit. Similarly, it is equally reasonable to argue that the co-existence of relatively high levels of IMF lending and high concessionality at the beginning of the 1980s was the result of monetary policies in industrial countries which forced commercial interest rates upwards, contributed to slow world economic growth and created balance of payments difficulties for many developing countries which led to their borrowing from the Fund.

For low-income countries drawing under the SAF and ESAF the Fund's rates have been maintained at a highly concessional level – 0.5 per cent at the end of the 1980s. Here again, of course, the price of loans is not the only issue. Low-income countries frequently face an availability constraint as far as commercial loans are concerned. The fact that they would find it difficult to service loans at commercial rates makes commercial lenders unprepared to lend. An increase in rates would merely make potential lenders even less prepared to lend to such countries since commercial loans would be perceived as increasing the financial burden on them unmanageably. Low-income countries may therefore encounter an inelastic or backward-bending supply curve of commercial lending with respect to the rate of interest.

Given that the interest rate on the use of SAF and ESAF loans has been maintained at a low level since their introduction, it would certainly be inappropriate to try and explain trends in the use of these facilities over time in terms of variations in the cost of drawings. For low-income countries using the SAF and ESAF other arguments in the demand function are more relevant.

In an attempt to gain further insight into what determines

*Figure 3.2* Fund lending to developing countries, 1980–92

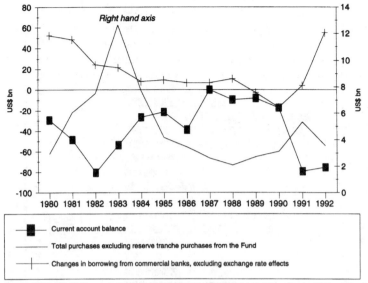

Source: IMF, *World Economic Outlook* and *International Financial Statistics*, *Bank of International Settlements*, Annual Reports, and national data.

borrowing from the Fund, Figures 3.2 and 3.3 provide information covering purchases on the Fund by developing countries, the size of their current account balance of payments deficits, the degree of concessionality on IMF resources excluding special facilities, and commercial lending to developing countries. Variations in conditionality are not included since there is no strong evidence of significant change in its strictness throughout the period covered (Killick, 1992). Perceptions amongst some potential borrowers may have changed but it is even more difficult to measure these.

No clear pattern can be discerned. There is the unsurprising tendency for purchases on the Fund to rise as the balance of payments deteriorates; this may be observed in both the early 1980s and early 1990s. As noted above, there also appears to be a positive relationship between purchases and the grant element of Fund lending, although in the early 1980s purchases peaked one year after the grant element; again such a lagged effect might be expected. In terms of the relationship between private borrowing and Fund purchases, this seemed to be negative in the early 1980s

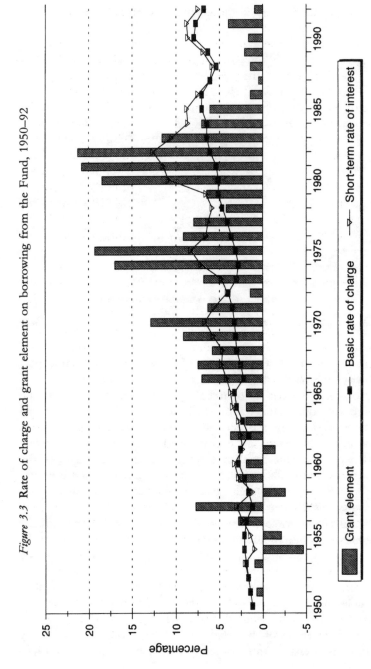

*Figure 3.3* Rate of charge and grant element on borrowing from the Fund, 1950–92

Grant element    Basic rate of charge    Short-term rate of interest

*Source:* Internal IMF estimates

but positive in the early 1990s, reinforcing an agnostic stance on the catalytic effect at this highly aggregated level.

Given the time series nature of the data, there are clearly too few observations to allow for econometric analysis. Moreover, the data are at too high a level of aggregation to provide a detailed picture. However, bearing in mind these restrictions it appears that it is the state of the balance of payments and the degree of concessionality which have the correct *a priori* signs. If, as claimed, conditionality did not vary significantly throughout the period, it follows that a given level of conditionality can co-exist with both relatively large and small amounts of credits from the Fund, although it may not be completely irrelevant that the largest level of credits followed a period spanning the end of the 1970s and the early 1980s when the general view is that conditionality was relatively relaxed. The evidence from the 1980s does not rule out the possibility that demand might be influenced by changes in conditionality.

How does the evidence of the 1980s and 1990s fit in with the public choice model of conditionality, which has it being used as a control variable to influence the demand for Fund resources? The model sees the Fund as seeking to increase both the amount of lending and the degree of conditionality, but sometimes having to compromise on conditionality in order to make borrowing from the Fund more attractive. The model claims to explain the pro-cyclical pattern of conditionality that some researchers identify. Here, when finance from other sources, in particular private capital markets, is available, the Fund is forced to lower conditionality in order to attract borrowers. However, when private finance is unavailable, the Fund has a captive market and raises conditionality.

A real problem in testing this hypothesis is to measure accurately the degree of conditionality. Although, as mentioned earlier, there are some indications of increasing conditionality during the 1980s, there are also indications of relaxation. On balance there is greater evidence of continuity than change. Time-invariant conditionality has coincided with strong, weak and modest market access for developing countries, as well as with large swings in IMF lending, whereas the public choice theory suggests that swings in conditionality would be used to stabilise IMF lending in response to the changing availability of private finance and changing balance of payments circumstances. Moreover, there is little evidence that IMF conditionality has been particularly strict in low-income countries as opposed to other countries, even though the public choice model predicts that it will be, since such countries do not have the opportunity to borrow elsewhere.

Variations in conditionality across countries may, however, be observed in terms of the Fund's temporary Systemic Transformation Facility (STF). Access to this facility requires countries 'to demonstrate [a] full commitment to co-operating with [the Fund] towards a goal of full fledged arrangements'. In this sense conditionality is relatively light. On the other hand, the Fund has also stated that a 'full commitment' may require 'significant policy actions as first steps . . . including prior actions' and conditionality may therefore not be as light as it initially seems. To the extent that it remains relatively light by comparison with other facilities, the explanation would not seem to be that the Fund is attempting to entice the economies of Central and Eastern Europe away from other creditors, since the private markets have not been keen to lend, but rather that political influences have dictated that special assistance should be offered to countries which are seen as being important strategically or that the economic assessment is that the process of transition does not lend itself to conventional Fund-supported programmes at an early stage.

Experience with the STF seems to be consistent with the more general point that variations in conditionality during the 1980s and 1990s have tended to occur more at the individual country level, driven by strategic and specific considerations, than over time as a way of manipulating the overall demand for Fund resources. Moreover, claims that the Fund induces the accelerated use of its resources by relaxing conditionality prior to potential quota increases do not hold up to the experience of the late 1980s, even though some researchers have claimed to find support for them in previous periods (Vaubel, 1991). Quota increases have, however, been found to be only loosely based on global economic factors and to be influenced by political considerations (Bird, 1987). Again, it may be noted that the STF was introduced some time after the Ninth General Review of Quotas had been approved. To the extent that this facility implied a relaxation in conditionality it post-dated rather than pre-dated the increase in quotas which is inconsistent with the public choice model. The increase in lending to Eastern Europe in 1990 occurred without any formal modifications to conditionality and within the context of then existing lending windows.

While the evidence seems to be largely at odds with a public choice explanation of conditionality, it remains interesting that the model claims that conditionality has the potential to be used as a control variable. The opportunity for influencing the use of Fund credit by means of varying conditionality should not be ignored.

If conditionality has not been used in this way, neither does it appear that the concessionary element has been used to entice countries to borrow from the Fund. Despite some evidence that Fund lending has varied with the degree of concessionality, the latter has been primarily affected by swings in market rates, and it is probably the credit rationing behaviour of the commercial sector that has exerted greater influence on the demand for IMF loans. The relationship between Fund borrowing and concessionality may not be causal. Moreover, the moral hazard argument based on the low cost of IMF finance is difficult to sustain where there have been such large and unpredictable variations in the grant element, and where, at times, it has been close to zero.

An interesting question emerges from this discussion: namely, why are Fund charges at the level that they are? The conventional answer is that they have to be related to market rates, especially in the case of resources borrowed by the Fund, in order to finance remuneration and induce creditor countries to make resources available to the Fund. However, this is an argument that may easily be overstated. If the demand for Fund loans is relatively inelastic with respect to the interest rate, is there any reason for believing that the supply is more elastic? Indeed, given the average *per capita* income of the Fund's major net creditors as compared with its net debtors, there might be a reasonable presumption that the demand elasticity would exceed the supply elasticity. Moreover, those countries that in the main finance the Fund's operations do not regard the provision of financial support as an investment which must generate a competitive rate of return. The return is more appropriately seen in terms of the global role that the Fund performs in attempting to eliminate balance of payments disequilibria and to stabilise the world economy, and in the private benefits which this indirectly confers on the Fund's creditors.

It is part of the revealed preference of creditor countries that they will provide concessional support for the Fund's lending; this is precisely what they do in the case of the subsidies attached to loans under the structural adjustment facilities.[53]

This position could change if the Fund grew significantly in size, and if the cost of running it rose dramatically. But at its existing scale of activities, it is difficult to see that the need to attract financial support is a convincing argument for increasing Fund charges to quasi-commercial levels. At the same time, if it is felt to be sensible to increase the lending role of the Fund in order to offset the market failures of private payments financing that were discussed in Chapter 2, then there could be an argument for lowering the cost of borrowing from the Fund in order to encourage users to translate potential into actual use.

## FUND LENDING RELATIVE TO NEED AND OTHER FINANCIAL FLOWS

We have seen in previous sections how IMF lending declined in the second half of the 1980s and some explanation of trends in lending has been offered. An additional perspective on Fund lending may, however, be provided by looking in more detail at its size in relation both to the size of balance of payments deficits that developing countries have been encountering and the size of other international financial flows. Reduced lending by the Fund would make more sense if the need for Fund support was declining because of improving balance of payments performance or because of increasing inflows from other sources.

Table 3.6 expresses purchases from the IMF as a percentage of current account deficits both for developing countries as a whole and for various regional sub-groups. For developing countries in general purchases in the period 1980–91 rarely rose much above 10 per cent. The exception is clearly the period 1983–5 when the percentage rose to as much as 40 per cent. It is interesting to note that for 1988–91 the figure fell from 14.5 per cent to only 5.7 per cent, suggesting again that for developing countries in general the Fund provided an insignificant amount of payments financing.

Broken down by regional sub-groups it may be seen that apart from 1980, 1982, 1985 and 1987, African economies generally received less balance of payments assistance from the IMF than did other developing countries. The largest IMF contribution to payments financing has been made in the case of developing countries in the Western Hemisphere, although for these countries there has been dramatic yearly variation. In 1980, for example, less than 1 per cent of their current account deficit was financed by IMF lending, whereas in 1983, 1984 and 1985 the percentages were 67.9 per cent, 432.4 per cent and 96.6 per cent respectively.

Table 3.6 Fund purchases as a percentage of current account balance of payments deficits

| | 1980 | 1981 | 1982 | 1983 | 1984 | 1985 | 1986 | 1987 | 1988 | 1989 | 1990 | 1991 |
|---|---|---|---|---|---|---|---|---|---|---|---|---|
| Developing countries | −9.4 | 13.9 | 10.0 | 30.8 | 40.0 | 22.7 | 9.3 | −24.3 | 14.5 | 11.8 | 10.4 | 5.7 |
| Africa | 30.5 | 7.1 | 11.0 | 14.5 | 15.3 | 84.1 | 6.2 | 7.1 | 4.2 | 8.2 | 5.9 | 5.4 |
| Asia | 8.5 | 15.0 | 13.6 | 25.0 | 37.4 | 7.5 | −21.1 | −2.6 | −1.8 | −28.0 | 3.2 | 13.7 |
| Europe | 4.8 | 12.8 | 40.8 | −18.5 | −12.8 | −4.6 | −3.0 | 0.0 | −2.1 | 1.1 | 1.6 | 19.9 |
| Middle East | 0.0 | −0.1 | −1.6 | 0.7 | 0.8 | 0.8 | 0.0 | 0.8 | 0.0 | 0.7 | 0.0 | 0.1 |
| Western Hemisphere | 0.8 | 1.1 | 4.2 | 67.9 | 432.4 | 96.6 | 9.2 | 16.8 | 12.5 | 28.0 | 61.4 | 6.5 |

Source: IMF, International Financial Statistics, January 1986, May 1992, and World Economic Outlook.

Thereafter the percentage fell to 9.2 per cent in 1986. By 1990 it was again over 60 per cent but fell to only 6.5 per cent in 1991. It is also interesting to note that, even though IMF lending to Central and Eastern Europe has received much attention at the beginning of the 1990s, it covered only 20 per cent of the current account deficit in 1991, at a time when drawings were relatively high.

These findings contrast, but are not inconsistent, with the conclusion drawn by Killick *et al.* (1992) that Fund credits are far from insignificant in relation to the balance of payments. In their study the sample is limited to developing countries which have negotiated an IMF-supported programme. For these countries over the period 1979–89 Fund credits were equivalent to about 30 per cent of the pre-existing current account deficit. Systemically, however, the fact remains that a much smaller proportion of the combined current account deficit of developing countries as a whole has been financed by the IMF.

An examination of net resource flows to developing countries confirms that IMF lending has in general also been insignificant in comparison with other forms of lending. As private flows diminished throughout the 1980s it was official development finance which increased partially to offset the decline; IMF lending did not fulfil this role and, indeed, made a negative net contribution. The share of ODF in total net resource flows to developing countries which had been about 30 per cent at the beginning of the 1980s had increased to over 60 per cent by the end. For countries in the Western Hemisphere the increase was from 10 per cent to over 50 per cent and for the low-income countries of sub-Saharan Africa from just over 65 per cent to about 96 per cent.

The same basic conclusion emerges as from earlier sections. Whether in relation to the size of balance of payments problems encountered or to the size of other lending, the IMF has only rarely made a significant quantitative contribution during the 1980s and early 1990s.

Of course, a small overall amount of IMF lending does not necessarily imply that it is inadequate. However, Killick *et al.* (1992) present evidence which is consistent with inadequacy. Examining IMF-supported programmes over the period 1979–89 they discover a clear and statistically significant relationship between the amount of IMF financing expressed as a percentage of the balance of payments current account deficit and the extent to which the programme is completed. The same is true when IMF credits are measured in relation to imports. Even when the data are adjusted for the cancellation and suspension of programmes, it is found that, whereas in the case of completed programmes the annualised mean use of IMF credits represented 48.3 per cent of the base period current account deficit and 16.9 per cent of base period imports, in the case of programmes that were not completed a mere 18.6 per cent of the current account deficit and 6.0 per cent of imports was covered by IMF credits. The finding that, relative to the balance of payments and imports, countries that did not complete their programmes received substantially smaller credits than those that did may be explained in a number of ways. But it certainly could suggest that inadequate financial support from the Fund contributed to the breakdown of programmes. Where more finance is available there seems to be a greater chance of success.

This finding in turn illustrates a potential connection between the Fund's financing and adjustment roles. Larger credits may effectively buy a greater commitment on the part of governments to conditionality and adjustment. In this context conditionality may be seen as a bargaining process between borrowing countries and the Fund. Public choice theory suggests that the Fund would like more of it and the borrowers less. Borrowers will, however, trade off conditionality against credits. Moreover, within developing countries those favouring IMF-type adjustment may be in a relatively weak bargaining position nationally, because of the politically sensitive costs of adjustment. In these circumstances having greater access to Fund credit will strengthen their hand and make it more probable that the programme will be completed. However, greater commitment may be induced not only by 'carrots' but also by 'sticks', and the Fund might seek to increase the level of commitment to and implementation of adjustment programmes by raising the costs of non-compliance in terms of reduced access to future credits, or increased conditionality, or both. The logic of this might be to have the Fund lending larger amounts under individual programmes but not engaging in an ongoing sequence of programmes.

However, perhaps we are getting a misleading picture by examining only the GRA and various special facilities such as the SAF and ESAF. Have developing countries made a heavy net use of their SDRs, and is this a significant source of balance of payments financing for them?

## THE SPECIAL DRAWING RIGHTS ACCOUNT: NET USE BY DEVELOPING COUNTRIES

Resources may be drawn from the Fund not only under the GRA and various special facilities but also under the SDR Account. During its first ten years the SDR represented a significant source of finance for developing countries which were strong net users of SDRs, and there was an active policy debate that a more formal link should be established between the allocation of SDRs and the provision of aid.[54] However, changes in the operation of the facility which raised the interest rate on net use to a market-related level largely eliminated the concessional element in the use of SDRs and this reduced their appeal to net users, although the fact that SDRs are essentially unconditional and have infinite maturity remained an attraction.[55]

Table 3.7 shows that since the last allocation in 1981 developing countries have made frequent but far from continuous net use of their SDRs. Indeed, they have made net acquisitions of SDRs in almost as many years as they made net use. The table suggests no clear relationship between SDR net use and the net use of Fund credit under the GRA and special facilities. But SDRs do not appear to act as a substitute for other Fund lending; indeed there is some evidence to support a complementary relationship. Faced with balance of payments difficulties, countries appear to draw finance from wherever they can. With low creditworthiness and limited international reserves they may turn to the Fund and in these circumstances will tend to use both GRA and SDR resources. Developing countries were net users of SDRs in 1983 when they also drew heavily on the Fund through its GRA facilities, and they acquired SDRs in 1985 at a time when the net use of Fund credit was zero. However, in 1988 when the net use of Fund credit was quite strongly negative, many developing countries made heavy use of their SDRs, with African countries reducing their holdings from SDR 300 m. to only SDR 89 m. and Western Hemisphere countries reducing their holdings from about SDR 1.28 bn (1,272 m.) to SDR 569 m. Of course, with a given stock of SDRs and no new allocations, developing countries cannot be continuous net users; they would simply run out of them. Instead, SDRs have to be used as an inventory that may be depleted in one year and replenished in another. Moreover, countries may wish to avoid reducing their holdings below a minimum

Table 3.7 Monetary authorities' holdings of SDRs: change on previous year (SDR m.)

| | 1980 | 1981 | 1982 | 1983 | 1984 | 1985 | 1986 | 1987 | 1988 | 1989 | 1990 | 1991 |
|---|---|---|---|---|---|---|---|---|---|---|---|---|
| World | −671 | 4,603 | 1,334 | −3,327 | 2,052 | 1,743 | 1,281 | 719 | −40 | 312 | −131 | 197 |
| Industrial countries | −436 | 3,051 | 2,153 | −2,571 | 1,853 | 1,527 | 1,233 | 341 | 1083 | 104 | −48 | −160 |
| Developing countries | −177 | 1,551 | −820 | −754 | 198 | 216 | 48 | 377 | −1,122 | 207 | −82 | 357 |
| Africa | −135 | 300 | −217 | −143 | −77 | 16 | 56 | 37 | −211 | 12 | −17 | 155 |
| Asia | −23 | 562 | −150 | −461 | 313 | 75 | −38 | −171 | −60 | −72 | 143 | −135 |
| Europe | −33 | 62 | −44 | 41 | −41 | 4 | 3 | 4 | 4 | 79 | −60 | 145 |
| Middle East | 151 | 210 | 434 | −144 | 126 | 10 | −132 | 75 | −151 | 219 | −341 | −38 |
| Western Hemisphere | −138 | 418 | −832 | −48 | −123 | 110 | 159 | 433 | −703 | −31 | 192 | 230 |

Source: IMF, International Financial Statistics.

level, and this may explain why two years of relatively heavy usage in 1982 and 1983 following new allocations in 1981 were themselves followed by four years when developing countries as a whole built up their SDR holdings from about SDR 2.9 bn (2,897 m.) to SDR 3.74 bn (3,736 m.). It is interesting to note that operational guidelines were issued to Fund staff in 1988 emphasising that countries should build up their reserves in order to ensure that they would be able to meet their commitments to the Fund; 1988, remember, was a year of relatively heavy net use of SDRs by developing countries.

There is little doubt that, in principle, the SDR offers one mechanism through which greater financial assistance could be made available to developing countries. Low-income countries in particular have low holdings of international reserves and have had some difficulties in repaying stand-by loans. Appropriate modifications to the SDR which would enable concessional allocations to be made to such countries could be of significant benefit to them without endangering global macroeconomic stability. The only constraint to such reform is the reluctance on the part of the Fund's major shareholders to allow the SDR to be used as a conduit for unrequited and permanent real resource transfers – a reluctance reinforced by the preference for conditional aid. At the same time, however, financial flows through the SDR facility do have some advantages from a donor's perspective by comparison with other forms of financial transfer.

## CONCLUDING REMARKS

The empirical evidence presented and analysed in this chapter enables a number of reasonably firm conclusions to be reached. First, although the Fund may provide significant amounts of finance to individual countries expressed as a percentage of their balance of payments current account deficits, it has provided only rather small amounts of finance relative to the payments deficits of developing countries as a group. At a time when developing countries have experienced severe payments difficulties the Fund has frequently been receiving more money back than it has been making available in the form of new loans. Against this general picture there have been significant swings in IMF lending over the period 1980–93, which in large measure reflect variations in the demand for Fund credit, although it remains difficult to predict which countries will and will not use Fund resources. The Fund cannot volunteer loans and has to respond to requests for assistance. However, it is not entirely passive. In principle it can attempt to influence the demand for its credits by modifying the 'price' it attaches to them in terms of conditionality and the interest rate charged. Although conditionality has been modified during the 1980s and 1990s, and although it is certainly not entirely uniform across all countries, there is little evidence to suggest that the Fund has manipulated conditionality with the prime motive of managing the overall use of its credit; the public choice explanation of Fund lending does not provide a satisfactory explanation. Yet the scope exists for re-examining the inter-relationship between conditionality and the use of credit; increased lending may enable more appropriate conditions to be made, and may induce greater commitment, greater compliance and greater success, which in the longer run reduce the demand for IMF credit.

Second, the Fund's lending role varies across countries. Where swings in Fund lending are observed these are often accounted for by relatively large programmes in a relatively small number of middle-income developing countries. Swings in the *amount* of Fund lending have therefore been more pronounced than swings in the *number* of programmes. Variations in the amount of Fund lending do not necessarily constitute a problem, since they may reflect the Fund correcting instabilities in market lending. Fund lending should have a 'public good' element to it. Low-income countries with low creditworthiness rely heavily on the Fund as a source of finance. Middle-income countries where creditworthiness varies make variable claims on the Fund. In some countries the Fund substitutes for other lenders, whereas in others it supports them by reducing the perceived risk of lending. Empirical evidence strongly confirms the notion that countries may graduate away from Fund borrowing through economic growth although the process of graduation may not always be smooth and irreversible. The identity of its clients may, however, create a slightly awkward situation in which the Fund is acting as a lending agency for only a sub-set of its membership, whereas in its systemic role it seeks to influence the behaviour of richer countries which do not borrow from it. It is unlikely that, with the degree of capital mobility in the world economy, the Fund will rediscover a lending role in industrial countries and it therefore needs to be asked whether it can

139

simultaneously sustain different sets of relationships with different members.

Variations in Fund lending during the 1980s and 1990s do not strongly confirm the public good role described above. Fund lending has often been pro-cyclical rather than counter-cyclical, and it is doubtful whether the Fund interprets its lending role in such public good terms. Instead, its overall contribution to international financing probably simply reflects the aggregate out-turn of a series of individual country-specific decisions. If this is the case, an important question is whether such an important international organisation should not possess a more well developed lending strategy.

The absence of a lending strategy is also suggested by a third empirical observation: large variations in the use of individual facilities such as the EFF and low usage of the CCFF and BSFF. Over the years the Fund has moved from a position where it claimed that no special facilities were required outside conventional stand-bys to one where there is an array of lending windows. The situation has indeed existed where Fund lending has been falling at the same time as the number of facilities has been increasing. For an outsider it is difficult to explain the changing use of individual facilities. Why was it, for example, that the EFF was so little used during 1984–9 when it is widely acknowledged that many developing countries, including the highly indebted ones, required structural adjustment? If EFF programmes generated poor results why was the facility reactivated in the late 1980s? Moreover, what is the purpose of having facilities such as the BSFF, and latterly the CCFF, which are so little used? Surely there must be scope for reappraisal and rationalisation. The Fund needs to define more clearly the pattern of international financing gaps as it sees them and design a more limited range of facilities which enables them to be filled. This would give Fund lending a much-needed sharper focus.

Fourth, the empirical evidence provides little support for the argument that there are severe moral hazard problems associated with Fund lending. Many countries that would be eligible to draw from the Fund choose to avoid it, and the Fund often carries significant spare lending capacity. Borrowing from the Fund is more appropriately seen as a last (and for some countries only) resort. Moreover, the claim that countries are encouraged to mismanage their economies in order to gain access to low-cost

Fund finance presupposes that Fund finance carries a significant concessionary element, and this is not always the case. However, the evidence also shows that poor compliance and programme failure do not significantly exclude countries from future access to Fund finance, and this may clearly weaken the incentive to make programmes work where at all possible. The view that Fund lending is temporary is empirically unsupportable, with evidence spanning many years and various sub-groups of developing countries showing that Fund involvement is often prolonged, if not sometimes quasi-permanent. In one sense this finding may be interpreted as enhancing the Fund's lending role, but it also raises serious questions concerning the Fund as an effective adjustment institution.

Fifth, the search for a catalysing role for Fund lending is at best only partially successful. Where econometric studies have found a statistically significant positive relationship between Fund and commercial lending this has usually been weak. At the same time, other research has suggested a negative relationship. Even at a highly aggregated level, Fund lending seems to move sometimes in positive and sometimes in negative association with private lending. This suggests that in determining private lending the involvement of the Fund, although certainly not irrelevant, is often dwarfed by other factors – a conclusion which is consistent with the theoretical discussion of the catalytic effect in Chapter 2. The concept of a strong and unambiguously positive catalytic effect is not supported by the evidence. This begs at least two further questions. First, if the catalytic effect cannot be relied upon, should the Fund take on a larger proportion of the financing role? And second, what can the Fund do to enhance its catalysing role?

Discussion of the Fund's catalysing role again demonstrates that Fund lending may have different purposes in different countries. Once more it appears that the Fund needs to identify more precisely the financing needs of its major clients and ensure that its own lending capacity matches these needs both in quantity and nature. Resolution of these issues will enable the Fund to present a more coherent and purposeful lending strategy than the one reflected by much of the evidence examined here.

# APPENDIX

## Developing Country Borrowing from the IMF: an Analysis of the Econometric Evidence

### THE THEORETICAL FRAMEWORK

The theoretical framework within which borrowing from the Fund may be analysed is relatively straightforward and may be derived from the fundamental trade-offs facing economies which encounter balance of payments deficits. The first trade-off is between adjustment and financing.[56] Here countries will seek to equate the marginal rate of transformation between the current and future sacrifice of expenditure with their marginal rate of substitution. Financing enables adjustment to be deferred, or undertaken more slowly, but it builds up future obligations and, as a result, future sacrifices. A key issue is therefore the country's social discount rate or social rate of time preference. This will depend not only on economic but also on political factors. Where a government perceives that there will be strong political resistance to current sacrifices of expenditure it will be encouraged, other things being constant, to substitute financing for adjustment.

Where a decision is made to finance a payments deficit a second trade-off becomes relevant. This relates to the various means through which financing may occur: the decumulation of international reserves, borrowing from other monetary authorities, borrowing from private international capital markets, or borrowing from the IMF (and other international financial institutions). This choice will be affected by the relative costs of the options. But the costs need to be defined quite broadly, and have to include not only the interest rate on the loan and its maturity but also other elements such as the loss of national sovereignty that might be seen as being associated with borrowing from the IMF.

Furthermore, it needs to be recognised that, while a choice between financing alternatives is involved, it will be a constrained choice. A country may opt for a specific amount of commercial credit only to find that its lack of creditworthiness precludes it from being able to attract the preferred amount of finance. The existence of a theoretical trade-off does not prevent a positive empirical relationship between the four financing sources. The trade-off will reveal itself in the proportionate relationship between the financing alternatives, and not in terms of the absolute amounts of finance.

With respect to IMF lending, the nature of the relationship between borrowing from the IMF and borrowing from private international capital markets is of particular interest since the Fund has itself frequently claimed that its loans have a strong 'catalytic effect' on private lending. In principle, however, IMF lending could be seen as a substitute for, rather than a complement to, private lending, and in this case a negative as opposed to a positive correlation coefficient would be expected.

It is only developing countries that have borrowed from the IMF during the 1980s and 1990s, assuming that Eastern European economies are classified as 'developing' for this purpose. This raises an interesting question about the relationship between the level of economic development and the demand for Fund credit. Our theoretical framework provides some insight in this respect. If it may be assumed that the level of economic development is inversely related to the costs of adjustment, because of relatively low response elasticities in developing countries and even lower elasticities in low-income countries, and positively related to creditworthiness, it would indeed follow that developing countries would be expected to borrow more from the IMF than developed countries. This would have Fund finance as an inferior good and there would be a negative income elasticity of demand for it beyond a certain level of development.

The theoretical considerations discussed in this section lead to a general model of the following form:

$$D^{IMF} = b_1 B + b_2 A^c + b_3 F^c \pm b_4 F^a - b_5 Y$$

where:

$D^{IMF}$ = drawings on the IMF

$B$ = the balance of payments deficit on current account

$A^c$ = the cost of balance of payments adjustment

$F^c$ = the cost of financing from sources other than the IMF

$F^a$ = the availability of financing from other sources

and $Y$ = national income *per capita*

A more specific form of this equation for estimation purposes might seek to reflect the targets that are set in most IMF-supported programmes. These include not only strengthening the balance of payments, but also reducing the burden of external debt and reducing the rate of inflation. The equation would then be:

$$D^{IMF} = b_1B + b_2DSR + b_3INF - b_4Y - b_5R \pm b_6P$$

where:

$DSR$ = the debt-service ratio
$INF$ = the rate of inflation
$R$   = the level of reserves
and $P$   = borrowing from private international financial markets

Our above analysis would lead us to expect that drawings on the Fund would be positively related to $B$, $DSR$, $INF$, and negatively related to $Y$ and $R$. The sign on $P$ is theoretically ambiguous.[57]

## THE EMPIRICAL EVIDENCE

As reported in the main text an early study by Bird and Orme (1981) based on the above equation met with only partial statistical and economic success. The model fitted the data well for 1976 with the estimated coefficients being consistent with *a priori* reasoning for all the explanatory variables except external debt, for which in any case the coefficient was not significantly different from zero. According to these results countries did seem to draw more resources from the Fund as their balance of payments deteriorated and as their rate of inflation accelerated, and seemed to draw less as they became richer. Private borrowing and borrowing from the IMF seemed to be complementary activities and, to this extent, some statistical support for a catalytic effect of IMF lending on private lending was discovered, although it was generally concluded that drawings from the Fund and the acceptance of the related conditionality were neither necessary nor sufficient conditions for access to private loans. However, the model broke down in 1977. Further analysis by Bird and Orme was unsuccessful in predicting which countries would and would not borrow from the Fund.

The general conclusion that econometric studies had yet to specify a satisfactory demand equation was supported by a further study by Cornelius (1987a) which focused on drawings by sub-Saharan African countries. Using pooled cross-section and time-series data he found that his equation, which was similar to the one used by Bird and Orme, provided a reasonable explanation of drawings over the period 1975–7 but not over the period 1981–3. For 1975–7 the results were generally consistent with the theoretical model although the coefficient on inflation was insignificant and negative. Also insignificant was the coefficient on the balance of payments; indeed for the 1981–3 period the sign was negative. Cornelius also found a negative coefficient on the private lending term over both sub-periods, thus again challenging the idea of a catalytic effect, at least for SSA countries. For the 1981-3 period he found that only one coefficient was significantly different from zero, and the coefficient of determination was only 0.54.

What do we find if we update these studies? Our data here are drawn from *International Financial Statistics* and *World Debt Tables* and include 235 observations of developing countries making purchases from the Fund during the inclusive period 1980–5. The data are annual. Re-running the Bird–Orme and Cornelius equations for 1980–5 using pooled cross-section/time-series data yields the following results where:

$$D^{IMF} \quad = \text{all purchases from the Fund}$$
$$IMP \quad = \text{the value of imports}$$
$$TRADE \quad = \text{the value of imports plus exports}$$
$$\text{and } CRE \quad = \text{new disbursements of debt creating private finance}$$

In the case of the Bird–Orme equation we get:

$$D^{IMF} = 13.0 - 0.45 DSR + 0.76 INF - 0.073 GNP$$
$$\phantom{D^{IMF} = } (0.32) \quad (-0.27) \quad (2.54) \quad (-3.29)$$

$$+ \ 30.0 \ IMP - 0.014 \ BOP/TRADE + 0.095 \ CRE/IMP$$
$$\phantom{+ \ } (10.1) \quad (-0.11) \quad\quad\quad\quad (4.88)$$
$$+ \ 0.14 \ RES/IMP$$
$$\phantom{+ \ } (1.48)$$

$$R^2 = 0.47 \qquad F = 29$$

and in the case of the Cornelius equation we get:

145

$$D^{IMF} = 53.0 + 0.0071 \ BOP + 0.78 \ DSR + 0.802 \ INF$$
$$\quad\quad (1.74) \quad (0.72) \quad\quad\quad (0.54) \quad\quad\quad (3.19)$$

$$-0.051 \ GNP + 8.57 \ IMP + 0.0080 \ CRE + 0.085 \ RES$$
$$(-2.59) \quad\quad\quad (2.02) \quad\quad (3.82) \quad\quad\quad (5.23)$$

$$R^2 = 0.554 \quad\quad F = 40$$

For SSA countries on their own, the Cornelius model yields the following results:

$$D^{IMF} = -43.0 + 0.023 \ BOP + 1.2 \ DSR + 2.19 \ INF$$
$$\quad\quad (-1.20) \quad\quad (0.50) \quad\quad (0.85) \quad\quad (4.30)$$

$$+ 0.0051 \ GNP - 17.0 \ IMP + 0.046 \ CRE + 0.28 \ RES$$
$$(0.07) \quad\quad\quad (-0.56) \quad\quad (1.96) \quad\quad\quad (3.02)$$

$$R^2 = 0.52 \quad\quad F = 5.5$$

For the Bird–Orme model the coefficients on inflation, GNP, imports and private flows are all significant at the 5 per cent level and have the correct sign. The balance of payments variable is negative but insignificant. The rather low $R^2$ reflects the large number of observations as compared with the original Bird–Orme study, although even in the rerun the results are strongly driven by a limited number of large drawings. Generally the inclusion of unscaled variables serves to give a misleadingly good fit for the regression equation under OLS techniques.

In the case of the rerun of the Cornelius model on the complete data set the results, at first sight, appear satisfactory. Only the coefficient on international reserves has the wrong sign, the $R^2$ is becoming respectable, and the $F$ value is highly significant. But again it appears to be the inclusion of unscaled variables that is producing an artificially 'satisfactory' fit; and even with these variables included the results for the SSA countries are statistically poor.

In an attempt to overcome the scaling and heteroscedasticity problem the Bird–Orme model was rerun using imports to scale drawings and including trade as the only size variable. Again covering 235 observations for the 1980–5 period this modified equation gave the following results:

$$D^{IMF}/IMP = 54.0 - 0.23 \ DSR + 0.22 \ INF$$
$$\quad\quad\quad (4.93) \quad\quad (-0.51) \quad\quad (2.59)$$

$$0.16 \; GNP - 1.02 \; TRADE + 0.094 \; BOP/TRADE$$
$$(-2.58) \qquad (-2.40) \qquad (2.63)$$

$$+ \; 0.018 \; CRE/IMP + 0.028 \; RES/IMP$$
$$(3.23) \qquad\qquad (1.11)$$

$$R^2 = 0.23 \qquad F = 9.73$$

In some ways the results are reassuring.[58] The coefficients on inflation, national income, the balance of payments and private financial flows are all statistically significant and have an economically sensible sign. The constant term is also positive and significant. Even the negative coefficient on the trade term does not create a problem, since it is quite plausible that as trade increases countries draw resources from the IMF which are equivalent to a smaller percentage of their imports. Only the coefficient on reserves has the wrong sign, but it is statistically insignificant. However, the removal of the unscaled variables dramatically lowers the coefficient of determination and the general explanatory power of the equation is poor, with over 75 per cent of the variations in drawings on the Fund remaining unexplained. The model is therefore of little use as a basis for prediction.

This general conclusion is confirmed by another study, again mentioned in the main text, which uses logit analysis in an attempt to differentiate between countries that do and do not enter into Fund-supported programmes in terms of specific economic characteristics (Joyce, 1992). This is a useful addition to our empirical evidence since in many respects IMF borrowing represents a discrete choice as to whether or not to turn to the Fund. The model estimated is again similar to the one discussed earlier, although Joyce includes the percentage growth in the central bank's holdings of domestic assets and the ratio of government expenditure to GDP as additional determinants, and he lags all the indicator variables by one period in order to avoid simultaneity bias. Studying a sample of 45 countries over 1980–4, which yields 225 annual observations and covers about two-thirds of all standbys agreed during the period, the results of maximum likelihood estimation of the equation suggest that countries borrowing from the Fund tend to have relatively rapid domestic credit growth and a relatively large government sector. The sign on the balance of payments is negative and significant; that on inflation is positive but insignificant; and those on reserves and *per capita* income are both negative and significant. The coefficient on private finance is negative but insignificant. However, what we again discover is that the general goodness-of-fit of the estimating equation is poor, as reflected by both the log likelihood statistic, which tests the null hypothesis that the coefficients of all the independent variables are equal to zero, and the log likelihood ratio, which measures the amount of uncertainty in the data which is explained by the variables in the equation. The model does not provide any basis upon which we can predict drawings from the Fund.

## INTERPRETATION, COMMENTARY AND CONCLUSIONS

What can we learn from the available empirical evidence about borrowing from the IMF? First, there is some suggestion that borrowing from the Fund is associated with a prior deterioration in the balance of payments but that Fund involvement strengthens the current account.[59] Second, countries tend to borrow more from the Fund if they are poor. All studies show a robustly negative coefficient on the measure of *per capita* national income. Economic growth and development can wean a country away from the Fund. It is therefore very important that the interface between development and the balance of payments is fully examined. If countries are to reduce their borrowing from the Fund they must secure economic growth. Fund-supported programmes therefore need to ensure that they are not hostile towards long-term economic growth.[60]

Third, the catalytic effect of Fund lending on private financial flows finds at best only weak support. Where a positive coefficient is discovered, it normally has a low value. By the same token, however, the evidence supports the notion that borrowing from the IMF and from private markets are more likely to be complementary rather than substitutable. Of course, the catalytic effect may take a slightly different form, and IMF lending may have a stronger impact on other official financial flows. While consistent with the concept of the Fund as a lender of last resort, the evidence suggests that other sources of finance still exist when countries borrow from it, even if access has been reduced over time. The Fund is therefore not the lender of last and only resort.

The main conclusion emerging from the empirical evidence is the resilience of the low explanatory power of the estimated equations. Something important is being omitted, or alternatively the equations have been mis-specified. In principle it might be that drawings on the Fund at one point in time reflect economic conditions in a previous time period. Moreover, perhaps variations in the nature and strictness of conditionality through time affect borrowing from the Fund. In practice, however, allowing for such factors does not improve the econometric results. Attempts to raise the coefficient of determination by lagging the independent variables, by fitting log linear versions of the equations, by allowing for exchange rate policy and by including dummy variables to capture changes in the degree of conditionality and other stochastic shocks, all failed to improve the fit.[61] Indeed, since conditionality is, in a sense, a fixed cost of borrowing from the Fund, one might expect it not to affect the marginal value of borrowing but only the discrete decision to borrow in the first place. It is surprising to find therefore that countries that turn to the Fund frequently borrow well below their access limits. It is also difficult to argue that the borrowing decision is affected by the relative cost of Fund finance, since over the period examined the interest charge on Fund loans stood fairly persistently below commercial rates and did not change significantly.

The remaining challenge is therefore to explain the residual variation which is left unexplained by economic characteristics. Observation of the pattern of Fund lending, as well as recent studies into the political economy of Fund conditionality, yield useful insights. Data on Fund lending reveal a strong recidivist tendency.[62] Countries that have borrowed once frequently borrow again and again. This implies some sort of 'borrower's threshold'. While they remain on one side of the threshold, countries will strongly resist borrowing from the Fund and will pursue a range of policies to avoid it. But once they have been forced into the Fund and have crossed the threshold the resistance to future borrowing is considerably reduced. Either the experience does not turn out to be as bad as they imagined it would be, or alternatively once the political price of using the Fund has been paid, the costs of future use are significantly reduced.

The political economy of Fund lending suggests that, quite apart from their economic performance, countries can negotiate different terms with the Fund depending on their political significance and their economic weight in the world economy.[63] Moreover, some countries may be able to muster more support amongst the membership of the Executive Board than others. The problem is that such factors are difficult to model formally and include in econometric estimation, but their exclusion may explain why demand functions which rely exclusively upon economic characteristics will leave much of the story untold. A country that, on the basis of its economic characteristics, might have been expected to borrow from the IMF may not do so, or may only borrow a small percentage of its quota, because it cannot negotiate effectively, while another country whose economic characteristics would imply a lower propensity to borrow from the Fund may borrow more than expected because of certain strategic advantages. Prior knowledge by a country of its bargaining position may well influence the decision to turn to the Fund in the first place. Since more countries might be expected to lack strategic significance and therefore bargaining strength than to have it, it is not surprising to find that studies based solely on economic characteristics tend to overestimate the number of countries that will borrow from the Fund.[64]

# 4

# IMF LENDING
## The way forward

What lessons may be drawn from the analysis and evidence presented in the previous chapters? What policy changes are appropriate in the context of IMF lending? What future does the IMF have as an international financing agency?

## LESSONS FROM THE PAST

The attack on the very existence of the Fund as an international lending institution, which is based on moral hazard arguments, has been seen to lack strong analytical or empirical support. While it may be true that some countries agree to adjustment programmes supported by the Fund without strong commitment and that the system allows or even encourages them to do so, there is no evidence that developing countries contrive to create balance of payments deficits in order to gain access to 'soft' Fund resources.[65] Indeed, much of the evidence points in precisely the opposite direction, with Fund finance perceived as being 'hard' both in terms of the low degree of concessionality it involves and the strictness of the conditionality that is attached to it. Models of the demand for Fund finance based on economic variables tend to over-predict Fund drawings. The picture still seems more accurately to be one of the Fund being regarded by potential clients as a lender of last resort – a source of finance to be avoided if at all possible.

Much stronger is the analytical and empirical support for the counter-view that, when it comes to the provision of balance of payments financing, markets fail in terms of efficiency and equity. This implies a role for the Fund in correcting the market's deficiencies. In some countries Fund lending will not be needed

151

and private international capital markets will suffice on their own. However, in others commercial lenders will not be present and exclusive reliance will be placed on official creditors such as the Fund. Low creditworthiness deters private lending, and international agencies with a different set of objectives, or simply a longer-term perspective, will be required if international finance is to be made available. In yet other countries the Fund will be present alongside both other official agencies and commercial lenders. The Fund can therefore operate not only in a way that offsets the inequities associated with commercial lending, but also in a way that moderates its inefficiencies. To the extent that the Fund induces greater commitment on the part of countries to pursue appropriate economic adjustment and, furthermore, augments the information and reduces the mis-information upon which lending decisions are made, it can enhance creditworthiness, correct informational asymmetries and strengthen the basis upon which country risk is assessed. In this way it can increase the trend availability of private international financing and minimise instabilities about the trend – instabilities that themselves create problems for the developing countries concerned.

The conclusion is that Article I (v) of the Fund's Articles of Agreement, which sees it as giving confidence to members by making its own resources available and also providing them 'with the opportunity to correct maladjustments in their balance of payments without resorting to measures destructive of national or international prosperity', remains analytically sound.

But at what level and in what form should Fund resources be made available? Again, with reference to the Articles of Agreement, what safeguards are 'adequate' for the purpose? Although Chapter 2 illustrates that the elements of a theory of optimal Fund lending may be fairly easily assembled, it was also seen to be exceedingly difficult to move from these broad concepts to any meaningful and objective operational model. This having been said, the reality of Fund lending is observable and the acceptability of the given state of affairs is therefore open to debate. Chapter 3 establishes two empirical facts. While the IMF has been a quantitatively significant source of finance to some individual developing countries at certain times, it has been quantitatively insignificant for developing countries as a whole when assessed against the size of their balance of payments deficits and financial flows from other sources. Moreover, for the second half of the 1980s net financial flows ran from developing countries to the Fund. The question one has to ask is whether this can be a satisfactory state of affairs, given the external financing gap that developing countries were encountering at the time, and given the economic adjustment policies that have had to be implemented and the economic and social consequences of these policies.

A strong argument can be made that the emphasis placed by the international financial regime upon adjustment relative to financing has been too great, with the result that global economic welfare has been adversely affected. By its own lending policies the Fund reinforced rather than neutralised this trend. As explained in Chapter 2, an external financing gap is an *ex ante* concept. By definition gaps cannot exist *ex post* because economic performance adjusts to conform with effective constraints. However, it remains a reasonable presumption that many developing countries have encountered, and are likely to continue to encounter, financing gaps such that economic development will be constrained by the availability of external finance. There are differences between developing countries but the evidence is that in general the Fund has failed to make a significant contribution towards closing this gap.

Relatively small amounts of its own financing would not, of course, necessarily imply a minimal effect on financing in aggregate if the Fund were to be instrumental in inducing additional private and official lending. The Fund has long emphasised this so-called catalytic effect in which its support acts as a catalyst for other financial provision. Indeed, the Fund has described its *central role* as 'supporting and *catalyzing other support* for members' economic adjustment efforts', (*IMF Press Release No. 92181*, November, 1992, italics added). However, the empirical evidence surveyed in Chapter 3 shows that, while specific instances of a significant catalytic effect may be identified, there is no strong general support for it. Correlation coefficients between Fund and other lending are usually found to be insignificant, with the sign of the coefficient as likely to be negative as positive. The negotiation of a Fund-backed programme seems often to be associated with a deterioration in the capital account of the balance of payments which suggests, at least superficially, that Fund support leads to an increase in net capital outflows. Of course, the precise causal connection is more difficult to pin down because of the problem of the counterfactual; what would have happened without Fund programmes? However, it remains the case that attempting to explain away negative results is never as satisfactory as having positive results in the first place, and the hard fact is that there is no strong positive empirical support for the catalytic effect of Fund lending.

How might this be explained? In principle there are two possibilities. The first is that while Fund-backed programmes do improve economic performance, the lending decisions of other lenders are not sufficiently sensitive to this improvement, so that the creditworthiness they 'perceive' is lower than the 'real' or 'actual' creditworthiness. Information is asymmetrical and, in this context, there is simply an informational failure that needs to be corrected. The empirical evidence, however, much more firmly supports a second explanation: that there is little reason to believe that IMF-supported programmes will significantly improve a country's economic performance either in the short or the long run. In these circumstances the existence of a programme supported by the Fund does nothing to increase 'real' creditworthiness; other lenders are accurately interpreting the information that is available and asymmetrical information is not a problem. Taken to the extreme the argument could be made that IMF involvement is not only an indicator of past economic difficulties but also a lead indicator of future problems, particularly given the recidivist tendencies of some of the Fund's clients. Here the involvement of the Fund stands as a proxy for declining creditworthiness, and in this case a negative catalytic effect is the expected and rational outcome.[66]

It is in respect of the catalytic effect that there is a strategically important connection between the Fund's impact on adjustment and other forms of financing, for it will be by increasing the effectiveness of IMF conditionality that the catalytic effect may be made more powerful. Enhancing the Fund as an adjustment institution is a prerequisite for enhancing it as an institution with a significant impact on international lending. If IMF programmes are ineffective in improving economic performance, and drive away rather than draw in additional capital, then the implication is that a much greater proportion of any given demand for balance of payments financing will have to be supplied by the Fund itself. The Fund then substitutes for other lenders.

A further connection between IMF conditionality and IMF lending also exists, since the latter reflects the interplay between the supply of and demand for Fund resources. While the Fund cannot lend more resources than it has available, it can certainly lend less, such that it is left with spare lending capacity. As noted in Chapter 3, it might be expected that, with other things being constant, the demand for Fund credits will be inversely related to the perceived 'costs' of IMF conditionality.

The 'costs' of conditionality comprise, first, the extent to which it deviates from what the government in question believes to be appropriate; second, the strictness of the conditionality in terms of the breadth and depth of the Fund's involvement in domestic economic policy as well as the scope for flexibility; and third, the financial cost of Fund resources and their 'grant element' or 'aid equivalent'. Inasmuch as it is possible to conceptualise an optimum quantity of Fund lending in the sense of maximising world economic welfare, it is important to examine both the supply and demand factors which determine actual Fund lending and cause the actual amount to deviate from the optimum amount. It is relevant to note therefore that, while the argument can be made that the Fund was systemically underlending during the latter half of the 1980s with developing countries placing excessive reliance on adjustment, it is also the case that it retained spare lending capacity. Was it then a tendency towards 'conditionality overkill' that so reduced demand as effectively to limit the Fund's financing role to a sub-optimal level? On this basis achieving a systemically optimal lending role for the Fund requires the appropriate combination of changes not only to its financing capacity but also to its stabilisation and adjustment conditionality. Making conditionality more effective and more cost-effective, and generally more acceptable, would in the short run increase the demand for Fund resources. In the long run, however, and with other determinants remaining unchanged, it would reduce the demand as balance of payments deficits were more effectively eliminated and as the perceived increase in the success rate of Fund-backed programmes had a positive effect on creditworthiness, thereby enhancing the Fund's catalysing role.

The discussion up to now covers considerations of efficiency. What about the equity of Fund lending? Discussion of equity raises a host of difficult and often impenetrable issues. At a superficial level the size and distribution of Fund lending may be observed and there can be some normalisation for the size and level of development of the borrowing economies. But beyond this it becomes problematical to assess Fund lending in relation to 'need', since this in turn requires an estimation of the size of individual financing gaps, creditworthiness and debt capacity. Moreover, while it is difficult enough to assess relative conditionality purely in terms of what is required of individual countries, it is much more difficult to relate conditionality to the adjustment *effort* needed and to the scope and costs of adjustment that are associated with Fund-backed programmes. Equity in terms of adjustment effort is therefore an awkward though nonetheless relevant operational concept.

However, a straightforward observation may be made concerning Fund lending and equity. It is clearly the case that private international capital markets will remain unwilling to provide substantial balance of payments financing to low-income countries. Moreover, given likely aid flows, it is also the case that such countries will continue to encounter financing problems. As the evidence presented in Chapter 1 illustrates it is low-income countries that experience the most acute balance of payments difficulties as a result of adverse movements in the terms of trade, export concentration and export instability, low creditworthiness, low holdings of international reserves, and high adjustment costs. It is quite consistent with its Articles of Agreement that the Fund should lend proportionately more balance of payments assistance to these countries. Indeed, it follows on from the analysis above that the Fund's catalytic role, particularly in terms of private financing, will be smaller in low-income countries and it will therefore be in these countries that the Fund has to substitute more fully as a source of finance.

Even with such assistance the outlook for many low-income countries is bleak. Without it prospects become yet more desperate. However, the analysis and empirical evidence examined here suggest that the Fund's general approach to the financing problems of low-income countries is not designed to maximise its beneficial impact. Low-income countries are perceived by the Fund as having economic problems with which the institution is not well designed to deal. Indeed, performance indicators confirm that the Fund's record in low-income countries is less good than elsewhere. Programmes perform less well, arrears are more common and the long-term nature of Fund involvement is more pronounced. Rather than wishing these problems away, the real challenge for the Fund is to design measures that will improve its performance in the poorest countries. In brief, and in the long run, this must imply a greater emphasis on economic growth since all the evidence shows a robust and strongly significant inverse relationship between Fund lending and the level of economic development. The dilemma is therefore that while low-income countries remain poor they will continue to turn to the Fund as a source of finance; however, for as long as Fund conditionality fails to foster economic growth, low-income countries will tend to remain poor. They will therefore be unsuccessful in graduating away from the Fund as a source of finance. While there is no single dimension involved in escaping from this vicious circle, there is a need for all those concerned to examine what needs to be done to raise the chances of lending progression and, in this sense, the Fund cannot circumvent its obligations any more than can governments and aid donors. To claim that the Fund is not a development agency is to obscure the issue.

While the evidence could be consistent with an explanation that accentuates deeper problems of economic mismanagement in poor countries, the credibility of this argument rests on being able to explain *trends* in performance. Is it that mismanagement has become more pronounced in recent years? If not, the suggestion must be that other factors are at work. An important part of the answer covers conditionality and the extent to which low-income countries may realistically turn around their balance of payments within the timeframe adopted by the Fund. But a second part of the answer covers both the size of access to Fund borrowing and its cost. If, for any given balance of payments problem, poorer countries need more access to non-private financing than do richer countries, access limits should surely reflect this. Moreover, if poorer countries are less able to pay near-commercial rates, should they not be granted a larger concessionary element on credit from the Fund? Finally, if it is accepted that in many instances, and especially where the intention is to minimise adverse effects on economic growth and living standards, balance of payments adjustment in developing countries will be a long-run phenomenon spread over many years, it follows that support from the IMF will also need to be long-run. In this regard the 'temporary' support envisaged in the Fund's Articles of Agreement needs to be interpreted as meaning 'non-permanent' rather than 'short-run'.

A reinterpretation of the Fund's Articles along these lines has direct consequences for its role as a lending institution. Basically two approaches present themselves. The first begins with a specific and predetermined lending capacity. Defining optimal Fund lending in terms of the size of individual credits and the maturity structure of lending then determines the maximum number of Fund-backed programmes at any one time. The second and superior alternative is to iterate from the definition of the Fund's lending role aggregated across all countries through to the amount of resources needed to perform it. In this respect it may be noted that the Fund's current lending capacity is vastly below that envisaged in many of the discussions which eventually resulted in the establishment of the Fund in the first place. The Keynes Plan, for example, saw an agency with a lending capacity equivalent to about 50 per cent of the value of world trade.

## A STRATEGY FOR THE FUTURE

Emerging from an analysis of the past it is possible to design a strategy for the future. The strategy comprises a limited number of basic components. But, while few in number, they are of fundamental importance and span both efficiency and equity.

The first component highlights the well-established inverse relationship between the availability of international financing and the speed of adjustment. The Fund should be seeking to work from the optimum adjustment path towards the availability of international financing, rather than beginning by imposing a financing constraint which then effectively defines the nature of adjustment. On the basis that it will generally be easier to manage aggregate demand downwards in the short run than to raise aggregate supply, rapid adjustment will have an in-built adverse effect on economic growth. However, economic growth that strengthens the trading sector lies at the heart of enabling countries to graduate away from the Fund. While the Fund has for a long time emphasised the need to address the balance of payments constraint on economic growth, and this is clearly important, it is equally important to recognise and address the economic growth constraint on the balance of payments. A shift of emphasis, and indeed a shift in philosophy, is required.[67]

The second component of the strategy recognises that working from adjustment towards external financing will have different consequences for the Fund in different countries. In low-income countries it will imply a relatively heavy financial involvement relative to country-specific variables, though light relative to the Fund's overall resources. However, an important part of the strategy must be to recognise the inter-temporal dimension of the Fund's involvement. If lower financial support now means less economic growth, it also means more balance of payments problems in the future and the need for continuing Fund support. The longevity of the Fund's involvement in many low-income countries aptly illustrates this. On the other hand, greater financial support now, in that it serves to alleviate financial constraints on economic development and facilitates economic growth, will strengthen the balance of payments in the future and will therefore reduce claims on Fund resources in the long run. The Fund needs to accommodate this long-term perspective.

To the extent that additional borrowing from the Fund places unmanageable financial burdens on low-income countries, as reflected in part by the increase in arrears, concessionality should become a central component of the Fund's strategy rather than remaining consigned to special facilities such as the ESAF. Here efficiency and equity coincide. The low absolute income of the poorest countries moderates the financial implications of such a strategy for the Fund.

A well-thought-out strategy for low-income countries might be expected to generate a positive response from aid donors which would reinforce it. For middle-income countries the Fund's catalysing role relates more to commercial lending. The strategy here should be to enhance the catalytic effect. However, enhancement again has to be placed in a long-term context which allows for credibility and time consistency. Exerting pressure on private international markets to lend on the basis of a false prospectus may have positive short-term effects but is likely to have negative effects in the long run as commercial lenders discount the Fund's involvement more heavily. The strategy must therefore be to view the Fund's catalysing function as a by-product of its conditionality role. Again the strategy must be to see financing and adjustment inputs as elements in an overall package of economic policies that must be mutually consistent. Financing is in the long run no substitute for effective and efficient economic adjustment, but in the short run it contributes towards making adjustment both more effective and more efficient. It is a significant weakness of much of the discussion of the Fund that it fails to draw out the two-way inter-relationship between Fund conditionality and Fund lending;

and it is important that in future the financing aspects of any adjustment programme are given greater weight and that the implications for Fund lending are fully considered.

A central component of a strategy for the future must concentrate on breaking out of vicious circles and into virtuous ones. In the vicious circle it is the lack of finance that dictates rapid adjustment; IMF conditionality is strict and hard, and is likely to involve economic and political costs. The high 'cost' of conditionality reduces the demand for Fund resources and delineates the Fund as a lender of last resort. The low success rate of IMF conditionality undermines any potential catalytic effect, with the result that the lack of finance is self-perpetuated.

In the virtuous circle finance does not so severely constrain the choice of adjustment path which may therefore be designed to minimise economic and political costs. In this way the commitment to adjustment and the feeling of ownership of programmes may be enhanced, with potentially beneficial consequences for their success rate. The very success of adjustment in the long run reduces the need for external finance to help deal with balance of payments deficits, but also increases the availability of commercial lending and the return of flight capital because of increased creditworthiness. The larger lending potential of the Fund, along with the reduced perceived 'cost' of Fund conditionality, increases the willingness of countries to turn to it earlier in the evolution of their balance of payments difficulties and, in the near term, raises the demand for Fund resources. It needs to be noted, therefore, that the spreading out of adjustment programmes will have both a direct and an indirect effect on the demand for Fund resources.

How can the Fund extricate itself from the vicious circle and replace it with the virtuous circle? It has to be by increasing the amount of financial assistance that is associated with Fund programmes, and in the short run this implies more lending by the Fund itself, since the greater availability of other finance requires the modifications to conditionality that are themselves only possible if more external finance is available initially. Greater financial support per programme could be achieved with existing Fund resources if the number of programmes supported by the Fund were to be reduced, although it also needs to be recognised that withdrawing support from one important middle-income country would enable the Fund to offer greater financial support to a significantly larger number of low-income countries. It could also clearly be achieved not by cutting the Fund's list of clients but by increasing the Fund's own resources. If one accepts the argument that systemically there has been too much emphasis on adjustment and too little on financing, then it follows that the latter direction is the better one to follow.

## ACTION POINTS TO IMPLEMENT THE STRATEGY

What actions would serve to implement these strategic changes? Without going into too much technical detail, it is possible to identify the sorts of policy changes that are required.

### Conditionality

The limited success of IMF conditionality could engender two extreme responses. The first would be to broaden its range and deepen its severity, on the basis that conditionality has not gone far enough to have a discernible impact. The second would be to remove or reduce conditionality, on the grounds that it has already gone too far and that removing or reducing it will at least allow countries to design their own adjustment programmes to which they will be more strongly committed. According to this view the poor track record of Fund-backed programmes makes them largely irrelevant.

Intermediate reforms such as those advocated by the author elsewhere (Killick *et al.*, 1984) set out to achieve a richer mix of policies which concentrate on the 'real economy'. The central focus of the reforms is to correct macroeconomic disequilibria as much as possible by strengthening aggregate supply rather than by reducing aggregate demand. Since increasing aggregate supply is not usually something that can be achieved quickly, many IMF programmes are likely to have to be phased over a larger number of years. Such reforms therefore need to be accompanied by reforms which cope with the increasing claims upon Fund resources.[68]

### The catalytic effect

This effect will be enhanced by any measures that strengthen the connection between IMF involvement in a country and its creditworthiness. However, strengthening this connection does not necessarily require an extension of IMF conditionality but rather an improvement in its economic effects. Indeed, it could be consistent with less strict conditionality if this in turn generated a stronger feeling of ownership and therefore more commitment to adjustment.[69] The greater use of pre-commitments by countries to involve the Fund more heavily in the design of programmes if their own efforts are unsuccessful and of contingency clauses by the Fund to avoid programme failure or the less transparent resort to waivers with their negative connotation, could help to nurture a positive effect on creditworthiness. However, in countries not at the margin of commercial creditworthiness the Fund needs to ensure that its own resources are adequate (along with other official lending) to fill financing gaps, and that its own policies towards adjustment do not needlessly limit the demand for its resources from countries with low creditworthiness because of excess conditionality.

## Access to Fund resources: conciseness and concessionality

Access to resources drawn from the Fund could be increased and streamlined in a number of ways. One possibility would be to abandon more completely a philosophy which involves quotas determining rights of access as well as obligations in terms of subscriptions. Even where subscriptions may be paid in domestic currency they imply additional fiscal burdens which accentuate domestic macroeconomic disequilibria. A rather different group of variables identifying need could then be used to determine access to Fund resources than would be used to determine subscriptions to the Fund, the latter being based on ability to pay. In effect, countries would have two quotas, one delineating their rights and the other their obligations. Provided that in aggregate the quotas ensured that the demand for Fund resources was matched by supply the redistributional role of the Fund should be adopted as part of its rationale.

The Fund could therefore offer more financial assistance to poor countries by pursuing a modified version of the 'small quotas' policy that it adopted in the 1950s and through which it sought to increase the borrowing capacity of some developing countries. The cost of increased lending by the Fund to developing countries would be covered by developed countries. While the

Fund maintains that recent quota increases should guarantee its liquidity over the next few years, there remains a strong case for replacing the process of quota reviews with a more automatic and ongoing mechanism for modifying the Fund's resources in the light of changing needs.

The streamlining of access could be achieved by rationalising the range of Fund facilities. These have evolved in large measure over the years to provide additional access to developing countries, but now incorporate facilities that are little used (BSFF, CCFF) and have overlapping purposes (ESAF, EFF).[70] Rather than necessarily reverting to one multi-purpose lending window, there would be something to be gained from a general reassessment of the Fund's lending facilities. The preferred outcome would depend on how far the reform of conditionality was to be taken. If financial programming was to remain at the heart of Fund conditionality, it might be appropriate to retain three lending windows: stand-bys, extended arrangements (including a policy framework component) and compensatory and contingency lending. Stand-bys would rely on conventional stabilisation measures. Extended arrangements would focus on longer-term structural adjustment but would still involve relatively strict conditionality, while compensatory and contingency lending would revert to the principles underpinning the CFF and would involve low conditionality but possibly with a pre-commitment element. If IMF conditionality were to be modified in a way that allowed countries a greater input in the design of adjustment, then the logic behind any distinction between compensatory and contingency lending on the one hand and other forms of lending on the other would become more blurred.

As things stand at present the reorganisation presented above would eliminate the main avenue through which concessionary assistance may be made available to the poorest developing countries, namely, the ESAF. However, two arguments may be made in defence of such a proposition. First, evidence suggests that ESAF programmes do not differ in any substantive way from EFF programmes. The need for longer-term support for structurally orientated programmes would continue to be provided by extended arrangements. Second, the logic of subsidisation should not be limited to a sub-set of Fund programmes, but for low-income countries should be extended to cover all Fund resources. Equity considerations form part of the justification for this, but efficiency arguments also support it. Arrears suggest that an 'IMF overhang' is particularly pronounced in low-income countries. Moreover, the severity of their balance of payments difficulties implies that it is precisely these countries that need to be provided with the additional incentive to turn to the Fund that would be created by more widespread concessionality. The constraint on the more wide-ranging use of subsidisation is financial, but it is a constraint that can fairly easily be overcome. First, the absolute amounts involved would be relatively small. Second, donors look-ing to offer greater assistance to low-income countries might be expected to find the subsidisation of IMF lending, with its in-built conditionality, a relatively attractive form of aid provision, although the issue of additionality would need to be addressed. Finally, extended subsidisation could be financed by further gold sales by the Fund, and, in this case, the direct fiscal and monetary costs of subsidisation for potential donors could be avoided; the costs would now be more indirect and would be in the form of the opportunity cost of gold sales.

## SDRs and financial assistance to developing countries

Additional Fund resources could also be made available via the SDR facility. The arguments for reactivating the allocation of SDRs and for distributing them to developing countries are compelling. In part the case is a positive one, but its strength also derives from the increasing irrelevance of the arguments that have traditionally been assembled against such a proposal.[71]

On the positive side the receipt of SDRs would help developing countries to close their external financing gaps. Moreover, addi-tional SDRs would initially increase their international reserves and would help to enhance their creditworthiness. During the 1980s many developing countries pursued policies designed to accumu-late reserves but these had high opportunity costs. For some, SDRs would complement commercial inflows, while for others they would be a substitute for them. Either way SDR allocation would help to alleviate existing debt problems in developing countries and would, in the future, provide them with a source of external finance which would not result in the build-up of external debt. More generally SDR allocation would represent one method of helping to reinstate positive financial flows to developing countries.

The precise nature of the distribution formula need not detain us here but could usefully build on the classification derived in earlier chapters which delineated countries by their likely claims on Fund resources. In effect and in proportionate terms the distribution of SDRs might be inversely related to the level of development and the availability of external finance from other sources. Data confirm that reserve inadequacy is particularly pronounced in low-income countries. For richer countries international reserve holdings as a proportion of trade are generally higher and in any case these countries have superior access to international finance.

The argument that developing countries would not benefit from receiving SDRs because their net use carries a market-related interest rate ignores the fact that, for many of these countries, it is the unavailability of finance rather than its cost that is the effective constraint. In any case the argument that SDRs need to carry an interest rate which is close to the market rate is much less strong in circumstances where the objective of establishing the SDR as the principal international reserve asset has in effect been abandoned. When initially established SDRs carried only a nominal interest rate. Even if the market-related rate were to be retained the principle of universal subsidisation could be extended to cover the net use of SDRs by developing countries.

Counter-arguments to reactivating SDR allocations and the SDR link do not stand up to close examination. The claim that SDR allocation is globally inflationary was always of quantitative insignificance, and remains so particularly in circumstances where the quantity of official reserves is less important than it used to be. Similarly the lack of conditionality associated with SDRs may be of less concern where the track record of IMF conditionality is itself viewed as rather unimpressive and where appropriate reforms may point in the direction of encouraging countries to design their own adjustment programmes. The existence of an adjustment programme could, of course, itself be made a prerequisite for the receipt of SDRs, and the greater availability of Fund finance generated by reactivated SDR allocations would provide additional incentives for developing countries to design such programmes. One idea would be to allow potential recipients of SDRs to choose their preferred combinations of conditionality and concessionality. Countries anxious to retain more sovereignty over the design of macroeconomic policy could opt to pay a higher rate of interest on the net use of SDRs, while others might prefer to trade off this freedom of policy manoeuvre against a lower rate of interest.

# THE POLITICAL ECONOMY OF IMPLEMENTING ACTION

Policy debate which ignores implementation is vacuous. There is little point in devising a strategy or a series of action points if the political economy of their implementation is not simultaneously considered. In the context of the IMF, action requires the support of the world's industrial economies. Without such support nothing will happen. With it, however, things can move far and fast as recent initiatives in the form of the Systemic Transformation Facility (STF) to assist Eastern and Central European economies, and in particular the former Soviet republics, have aptly demonstrated.

It may be unrealistic to assume that similar momentum could be generated behind proposals to assist developing countries, but it is realistic to consider whether and how the acquiescence of the developed world might be fostered. As a way of analysing this let us assume that in determining their 'position' the developed countries will endeavour to determine the implications of change (as opposed to no change) for the 'system' as a whole but in particular for their own situation within the system; they will attempt to assess both the direct and indirect private costs and benefits. These are not things that can be measured accurately and objectively, but which depend on the implicit 'model' used to explain how the system works. Modifications in the underlying model used will cause positions to change.

Since the 1980s there have been important changes in both the political and the economic model which determines the negotiating position of the developed countries in international financial matters. On the political front the thawing of the Cold War enables issues of international redistribution to escape from an East–West framework. Since no equally clear political model has replaced it, this should provide an opportunity for economic factors to have relatively greater influence.

On the economic front there has been a retreat from the neo-classical paradigm with its strong (and sometimes almost exclusive) emphasis on markets. This retreat has brought with it a greater recognition of the possibility (probability) of market failure and of the related need for governmental intervention. Illustrative of this shift in attitude is the increasing emphasis on macro-economic co-ordination. At the same time, difficulties that have

166

been encountered in making progress at a regional level may be expected to reawaken interest in international co-operation – something that is further endorsed by worries about unfettered exchange rate flexibility and protectionist commercial policy.

In the context of the 'new' economic model it is much more likely that the strategy described above will be viewed as conferring systemic economic (and political) benefits on the developed countries by means of increasing international stability. Fears in the developed countries that enhanced IMF lending will be globally inflationary are significantly weaker in an environment where inflation is globally low, unemployment is high, and economic stagnation is perceived as the principal problem. Moreover, the 'position' that more lending by the IMF will be financially de-stabilising derives more easily from the monetarist perspective of the 1980s than the more eclectic approach which now seems to be in vogue, and carries little weight where IMF lending represents such a small component of international liquidity.

Furthermore, although developing countries may worry that lending to Eastern Europe will crowd them out of IMF lending, it should also reassure them in the sense that the Fund's shareholders have, through their actions, demonstrated that they implicitly accept the key role that external financing plays in securing prolonged economic development. Again this implies a significant shift in the underlying economic model that is being used to guide decisions.

While the developing countries will be anxious that greater IMF lending does not simply replace other forms of external financing with the result that there is no net gain and even a net reduction, the developed countries might be expected to see the reforms analysed above as carrying few direct costs for them and therefore be prepared to accept them as additional forms of assistance. Allocating SDRs to developing countries, for example, carries a potential real resource cost for countries supplying the exports that the developing countries may now finance, but it does not carry the domestic fiscal and monetary costs of other forms of aid provision.[72]

The evidence suggests, and the impression is given by discussions with Fund staff, that there is an albeit gradual but nonetheless serious attempt in process to modify the Fund's operations in the light of changing circumstances as reflected by the shift away from stand-bys and towards extended arrangements and structural lending, the consequence of which has been to lengthen the average maturity of Fund lending to about eight years as compared with about five years in the mid-1980s.

While it has been argued here that such changes are appropriate they will continue to highlight the overlap between the Fund and the World Bank. Greater emphasis on economic growth and the protection of vulnerable groups brings the Fund into the Bank's traditional territory just as structural lending brought the Bank into areas previously associated with the Fund. There are various ways in which this overlap may be handled which have been spelt out and analysed in detail elsewhere by the author (Bird, 1993a, 1993b). Amongst them one possibility would be to take the opportunity provided by a move back towards global exchange rate management and the international co-ordination of macroeconomic policy to shift the Fund once again into a systemic, supervisory and regulatory role and to shift the role of policy-based lending to developing countries more heavily or even exclusively towards the Bank. Such a shift would clearly dramatically reduce or even curtail lending by the Fund.

Realistically, however, it is difficult to envisage major institutional reform of such proportions. Institutional structures experience strong elements of hysteresis. Moreover, evidence from the past suggests that it is also unrealistic to expect that dramatic discrete changes in IMF lending to the developing countries will occur. But at the same time the particular confluence of economic and political factors that exists as we move through the 1990s suggests that the environment may be more conducive to implementing the strategy and actions analysed here than it has been for some years. The opportunity exists for the Fund to take an important step forward in its relationship with the developing world.

# NOTES

1 For a brief discussion of the operation and weaknesses of the Bretton Woods system see Bird (1985).
2 Some analysts have argued strongly that the adequacy of international reserves was never as important an issue as it was thought to be in the 1960s, and that the so-called Triffin dilemma associated with the key role of the dollar was not a dilemma at all, see Chrystal (1990). For a rather different perspective, see Bird (1985).
3 For a further discussion of the failure to establish a NIEO, see Bird (1988: Chap. 12). It may be noted that as yet an effective debtors' cartel has proved as illusory as a commodity cartel.
4 Public choice theorists would no doubt have characterised the Fund as desperately searching around for an alternative role in order to justify its existence and the relatively high salaries it paid its staff (Vaubel, 1991).
5 Khan and Knight (1988), for example, on the basis of a study of 34 developing countries have identified a strong positive correlation between the availability of imports and export volumes. In addition to this, Otani and Villanueva (1990) find strong quantitative support for the view that export performance has a dominant influence on economic growth in developing countries.
6 Love (1990) estimates instability based on linear, logarithmic and moving average trends across a sample of 65 developing countries as well as for a more restricted sample of 58 developing countries. Taking the larger sample and a logarithmic trend, for example, he finds that the mean instability index rose from 0.155 in 1960–71, to 0.256 in 1972–84. The observed increase in export instability for developing countries in both Asia and the Western Hemisphere is perhaps particularly interesting since it is frequently assumed that export diversification and a move away from primary products and into manufactures will solve the problem. Certainly the conventional view has been that poorer developing countries with a higher concentration on primary products will be more vulnerable to export instability. Helleiner (1983a), for example, notes that during the 1970s it was the least developed countries that experienced the highest levels

169

of instability in their terms of trade, the purchasing power of their exports and their import volume.

7 See Bird (1978) and Chrystal (1990) for a review of the theory of the demand for reserves. Bird also relates this theory to developing countries.

8 For a further exposition of the problems of adjustment in developing countries, see Helleiner (1986). The argument that adjustment has a particularly adverse effect on the poor is to be found in Helleiner (1987). Discussion of the political costs of stabilisation may be found in Bienen and Gersovitz (1986), Haggard (1985), Nelson (1984; 1989), Sidell (1988), Haggard and Kaufman (1989), Kaufman (1988) and Stallings and Kaufman (1989).

9 See, for example, Nowzad (1989), who argues that 'the Fund is not (as some have suggested) involved for any particular partisan reason, such as bailing out the banks, or enforcing the policies of creditor countries' (p. 120).

10 For a clear critical appraisal of the Fund's involvement in highly indebted countries, which develops many of the issues introduced here, see Sachs (1989a, 1989b) and Edwards (1989).

11 See, for example, Maynard and van Ryckeghem (1976). There is now a huge literature dealing with the question of devaluation in developing countries, and a number of reviews of this literature. For a brief review of the reviews, see Bird (1990a) which also provides a broader survey of developing countries within the international financial regime. Whatever the problems with devaluation, it needs to be assessed against the alternatives. Here there is considerable evidence to suggest that devaluation is often the least-cost option and that there are significant costs associated with maintaining a disequilibrium real exchange rate (Edwards, 1988). A related criticism of the Fund, however, is that, once having encouraged the establishment of an equilibrium real exchange rate, it has done insufficient to encourage the maintenance of that rate through the use of some form of sliding parity.

12 See Edwards (1989) for a presentation of the evidence. He examines 34 upper tranche programmes approved in 1983 with developing countries, the vast majority of which had serious debt problems. He finds that, 'almost every programme contained credit ceilings and a devaluation component' (p. 32), arguing that 'this contrasts sharply with previous Fund programmes'. Edwards uses two methods for assessing the programmes. The first is a simple before and after comparison. He finds that 'on average, the current account improved somewhat while inflation increased quite significantly. With respect to output growth, after a steep reduction in 1983, there was a small improvement in 1984 and 1985.' He notes that 'countries that did not have Fund programmes also experienced major current-account improvements' (p. 34). Second, he compares targets and outcomes. His results show a rather low rate of compliance, both in absolute terms and by comparison with previous periods.

Sachs (1989a) argues that 'the evidence presented in the IMF's

1988 review of conditionality also suggests that, since 1983, the rate of compliance has been decreasing sharply, down to less than one-third compliance with programme performance criteria in the most recent years' (p. 107). Edwards notes that 'a serious consequence of the low rate of compliance has been that in recent years there has been a significant increase in the number of programmes that have been interrupted as well as in the number of waivers approved by the Fund' (p. 36).

Of course, the evaluation of IMF-supported programmes is fraught with methodological problems. Khan (1991) reviews these problems in considerable detail but, on the basis of his own tests covering Fund-supported programmes during 1973–88, concludes that they have generally been associated with an improvement in the balance of payments. The improvement was particularly marked for the current account, where the implementation of a programme led, on average, to about a 1 percentage point improvement in the ratio of the current account to GDP. In contrast to Edwards and Sachs, Khan claims that the evidence indicates that Fund-supported programmes have been more effective in improving the external balance in the 1980s than they were in the 1970s. Indeed he suggests that while 'we do not as yet have the final word on the effects of programmes . . . it does appear that these effects are more positive than has previously been reported' (p. 224). Although Khan does not take into account the degree of policy implementation (since he maintains that it is 'not easy' to do so), he suggests that exclusion of this factor may lead to an underestimation of effectiveness. 'Had the tests been restricted to only those countries that successfully implemented the recommended policies, it is conceivable that an even more positive picture would emerge' (p. 223). This view again differs from that of Edwards and Sachs, who argue that the degree of compliance should itself constitute a measure of effectiveness. Clearly there is still some way to go before a broad consensus emerges on the effectiveness of Fund-supported programmes.

13 Thus Nowzad (1989) emphasises how the conventional relationship between the Fund and the banks was reversed in the early years of the debt crisis.

14 See Killick (1989a) and Edwards (1989) for a clear statement of this view. Some elements of the newer open economy theories could be of particular relevance to the IMF. The use of waivers, precommitments and contingency lending could all have implications for the Fund's reputation and the credibility of the policies it supports. Edwards (1989) argues that 'time consistent arguments can be used to provide a firm theoretical justification for conditionality' (p. 21). Sachs (1989a) maintains that, even on its own terms, the Fund's underlying theoretical model is rudimentary in assuming a fixed velocity of circulation, crude links between economic growth and imports and a fixed incremental capital output ratio. Vines (1990) complains that even the more sophisticated IMF models make assumptions and contain omissions which have important ramifications for the policy

171

conclusions which emerge, basically serving to understate the negative output effects of Fund-supported programmes. Killick's critique is broader and suggests that the Fund has not only overlooked theoretical advances in terms of the likely effects of demand-side measures, but has also adopted a simplistic approach to analysing the supply side and the role of the state in developing economies.

15 See, for instance, IMF (1986), where it is suggested that 'if Fund-supported programmes imply that specific income classes (and in particular the poor) inevitably bear the brunt of the economic costs involved, then those programmes would be both less acceptable and, in the long run, less effective than the available alternatives' (p. 1).

16 For examples of this literature, see Killick *et al.* (1984) and Helleiner (1983b). Helleiner identifies many differences between African and Latin American economies, characterising the typical African economy as 'smaller, poorer, more trade-dependent, less urbanised, and less socially stratified than its Latin-American counterpart. Its agricultural sector weighs more heavily in overall output and is based much more upon small-holder production; the urban work force is not only relatively smaller and politically weaker, but also usually enjoys close links to rural families. Its financial institutions are weaker and more rudimentary. Despite the dramatic acceleration and education programmes in the post-independence period, levels of literacy and educational achievement are still relatively low in Africa. The ability to govern is limited by severe shortages of appropriate skills, not least in the area of economic analysis' (p. 10). He notes that, 'these intercontinental differences play upon the politics and economics of alternative stabilisation or adjustment programmes.' Killick (1989b) points out that it is important not to lose sight of the fact that not all low-income countries are to be found in Africa. Those in Asia and the Pacific also encounter similarly severe problems.

17 See Loxley (1984) for this conclusion. Zulu and Nsouli (1985) also discover a rather poor record for Fund-supported programmes in Africa. For a general review of low-income countries within the international monetary system, see Bird (1983).

18 Goreux (1989) presents evidence to show that the Fund has had outstanding credit in some low-income countries for between 10 and 29 years.

19 Goreux (1989) states that, at the end of the 1980s, 4 countries accounted for 80 per cent of total arrears due to the Fund. Helleiner (1983a) shows that, in earlier periods, the problem of arrears was heavily concentrated in low-income countries. Thus of 32 countries in arrears on their external payments in 1981, 20 were African.

20 The argument has perhaps been most strongly stated by Sachs (1989a).

21 For a discussion of this see Bird (1983), which also contains a broad review of the IMF and the developing countries.

22 For a review of such proposals, see Corden (1988). It should be noted that the Brady Plan might be expected to have a negative effect on resource transfers, since a given amount of finance directed towards

debt reduction will have a smaller impact on current flows than would a similar amount of new lending.

23 For an analysis of the optimal blend of financing and adjustment see Bird (1978), where it is concluded that this occurs where the community's marginal rate of substitution between current and future expenditure equals the marginal rate of transformation between the sacrifice of current expenditure (adjustment) and future expenditure (financing).

24 Vaubel (1986) provides an interesting discussion of the public choice aspects of international agencies, which develops many of these ideas more fully. A later paper (Vaubel, 1991) undertakes a detailed analysis of the Fund's activities along public choice lines.

25 The economics of shock versus gradualist policies is still poorly developed, but see, for example, Edwards and Montiel (1989). One difficulty is that much depends on behavioural and expectational changes which are difficult to model.

26 A strong counter-claim, namely that the system needs a lender of the last resort in order to ensure stability, can of course be made. On this specific issue see Griffith-Jones and Lipton (1984).

27 The 'mutuality of interests' argument formed a central element in the Brandt Commission's justification for enhanced international resource transfers (Brandt, 1980).

28 This argument is not infrequently presented in oral discussions with Fund staff. However, it has been expressed in writing by a former senior member of the Fund's staff, see Finch (1989). Also relevant to this debate is a series of papers by Guitian (1991a, 1991b, 1991c).

29 Killick (1989a) provides a comprehensive review of the re-emergence of the neo-classical paradigm with its implication of a reduced role for the state, and illustrates how proponents of the market mechanism have been unable to counter the concern over market failure.

30 This argument was strongly made in Dell and Lawrence (1980) but also finds support from empirical studies undertaken within the Fund itself (Khan and Knight, 1983). See Killick *et al.* (1992) for further evidence to suggest that exogenous factors are important in explaining why countries turn to the Fund.

31 Bird (1989), for example, provides various suggestions concerning the variables that should be consulted by the banks in assessing risk.

32 There is a potentially important inter-temporal dimension that also needs to be borne in mind. Closing a financing gap by means of compressing imports may mean that the financing gap is made wider in the long run since today's imports may be significant inputs for tomorrow's exports. Khan and Knight (1988) provide empirical support for such concern.

33 Bird (1978) provides a review of the theory of reserve adequacy. This has not altered very much in recent years, as the issue of reserve adequacy has generally appeared less relevant in a world of flexible exchange rates and private international liquidity. Recent contributions to the empirical literature include Edwards (1983), Chrystal (1990) and Frenkel (1984).

34 Williamson (1983) also provides an albeit loose analytical framework for considering the optimum quantity of Fund lending and its related need for resources, based on estimating financing gaps within the international financial regime.

35 Edwards (1989) provides a brief insight into what contribution might be made by recent advances in macroeconomic theory towards analysing the role of the Fund.

36 A slightly different defence of low conditionality lending to low-income countries has been made on the grounds that such countries have not enjoyed the growth in international reserves associated with the appreciation in the price of gold, since their holdings of gold are low. Richer countries with more substantial gold holdings have, on the other hand, experienced an increase in unconditional international liquidity (Brodsky and Sampson, 1981). Public choice theory indicates that the Fund would prefer high conditionality on the grounds that this increases its power and influence. However, low conditionality may increase the demand for Fund resources and this is another component of the Fund's objective function according to public choice analysis, which as a result must remain somewhat agnostic on the Fund's preferences as regards the level of conditionality.

37 Bird (1985) provides a brief introduction to the SDR, discussing its advantages and disadvantages as compared with other reserve assets. For an analysis of the link proposal, see, for example, Bird (1978). For more recent discussion of the SDR, which concentrates on enhancing its attractiveness see Coats (1990), and for an analysis of international resource transfers associated with the SDR, see Coats *et al.* (1990). Further useful analysis of the SDR can be found in von Furstenberg (1984).

38 Chrystal (1978) and Lal (1980) give an outspoken critique of the SDR. See also Chrystal (1990) for the argument that reserve adequacy was never the problem it was assumed to be during the 1960s and that the SDR facility was not needed.

39 There have been reconstitution requirements associated with SDRs which limit the extent that they can be used in the long run, Bird (1982), Williamson (1984).

40 Bird (1978) discusses the 'informal' link and the benefits that developing countries have derived from the SDR facility in terms of resource transfers. On the latter issue see also Bird (1979), (1981b), (1983).

41 As noted earlier, the Fund's public-good nature is interpreted by some as not requiring it to lend and therefore not requiring finance.

42 Kenen (1985) describes a more involved scheme based on SDRs.

43 Brodsky and Sampson (1981) make a strong case for the further use of Fund gold to finance a development account. Bird (1981a) has similarly shown how IMF gold sales could be used to finance the expanded use of subsidies, and how the operations of a substitution account could be tied to gold decumulation by the Fund.

44 For an analysis of the CFF and a discussion of the 1983 review see

Bird (1987) and Dell (1985). Bird (1990a) provides an early critical assessment of the CCFF.

45 It may be remembered that the CFF was designed to provide finance at low conditionality to help countries cope with the balance of payments implications of temporary export shortfalls. The CFF provided finance largely unconnected with adjustment. The logic of the CFF was related to export instability rather than any deterioration in the terms of trade in developing countries. While many have continued to experience an adverse movement in their terms of trade the available evidence also continues to suggest that there is significant instability about this trend.

46 Further confirmation of the longevity of the Fund's involvement with developing countries is contained in Killick *et al.* (1992) and Conway (1991). From a sample of 53 countries that participated in stand-by agreements with the Fund during 1977–86, 30 had Fund agreements in operation for 5 years or more.

47 In June 1990 total IMF quotas stood at just over SDR 90 bn (90,133 m.) for a membership of 151 countries, whereas in 1983 they had stood at just under SDR 90 bn (89,236 m.) for a membership of 146 countries.

48 In 1988, for example, only Mexico and Sudan were using GRA credit in excess of 300 per cent of their IMF quota. On average the figure for developing countries was 67.1 per cent.

49 This confirms the persistence of Fund involvement in many developing countries that was noted earlier.

50 Changes in the nature and importance of the catalytic effect could result from the Fund's changing clientele.

51 Bird and Helwege (1994) identify some resurgence in commercial lending to Latin America at the beginning of the 1990s while net credits with the IMF were still negative. This may, of course, imply that there is a lagged effect of Fund involvement on creditworthiness and therefore private lending, but econometric studies generally find that a lag structure does not improve the significance of the correlation coefficient. The somewhat ambiguous relationship between IMF and commercial lending via the impact of Fund involvement on creditworthiness remains in some ways unsurprising. Interviews with bankers in both the US and UK during the mid-1980s showed that, while the majority included IMF involvement in their credit assessment matrices, some included it with a positive sign, taking it as an indication that economic management and performance would improve, while others included it with a negative sign, taking it as an indication of severe economic difficulties.

52 In the Appendix to this chapter it is suggested that borrowing from the Fund is subject to a threshold. While countries remain on one side of the threshold they will strongly resist borrowing from the Fund and will pursue a range of policies to avoid it. But once they have been forced into the Fund and have passed over the threshold the resistance to future borrowing is considerably reduced. Either the experience does not turn out to be as bad as they imagined it would be, or

alternatively, once the political price of using the Fund has been paid, the costs of future use are significantly reduced. It is also argued that the different political and strategic importance of countries influences their borrowing from the Fund, and that the fact that most developing countries lack such importance explains why models of borrowing from the Fund tend towards over-predicting the number of Fund loans.

53  We may note that the shares of ordinary and borrowed resources in financing Fund assistance have altered quite significantly during the period 1980–90. Borrowed resources accounted for 40 per cent of the total in 1980 rising to 44 per cent in 1985 and 1986. However the share of borrowed resources fell to only 33 per cent in 1989, even though there had been no increase in quotas since 1983. This provides another indicator of the spare lending capacity held by the Fund at this time, and to this extent market-related rates were not needed in order to attract resources to the Fund.

54  For a summary of the principal issues in the debate see Bird (1978). Other studies of the SDR which examine its potential as a means of channelling real resource flows to developing countries include Bird (1979, 1981b, 1982). Williamson (1984) argues that there are both systemic and specific arguments for reactivating the SDR and in a recent paper (Williamson, 1992) again makes a case for its use to assist developing countries. Renewed interest in the SDR is being shown in a world economic environment where some believe there is a shortage of international liquidity (Coats, 1990; Coats *et al.*, 1990).

55  Bird (1981b), however, argues that, even with a market-related rate on net use, SDRs could provide useful assistance to developing countries. In any case, and in principle, subsidies could be used in order to reduce the cost of net use to all developing countries or to a sub-set of them.

56  For a more formal presentation of the theory discussed in this section (Bird, 1988a, Chapter 1, 'The Mix Between Adjustment and Financing').

57  There is, of course, the problem of two-way causation. While it might be expected that countries with poor balance of payments and inflation performance would be more inclined to use the Fund, use of Fund finance and exposure to the related conditionality might be expected to strengthen the balance of payments and reduce the rate of inflation. This will not be a problem if IMF-backed policies take 12 months or more to have an impact, and in any case, a lagged version of the equation may be tested.

58  We recognise that drawings will be truncated at zero and that this violates one of the assumptions underlying OLS estimation. Even so the technique is sufficiently robust for our purposes. OLS is not uncommonly used with variables such as height and weight although these are clearly positive.

59  This suggestion is further supported by the results of an as yet unpublished IMF paper which finds that both domestic economic mismanagement, in the form of excessive credit creation, and a

deterioration in the terms of trade cause balance of payments problems prior to participation in an IMF-backed programme (Conway, 1991).

60 There is some evidence that IMF-supported programmes have a negative effect on economic growth at least in the short run (Khan, 1990; Killick *et al.*, 1992).

61 Devaluation of the exchange rate is unlikely to be avoided by turning to the Fund since the clear majority of IMF-supported programmes incorporate devaluation as a precondition or a performance criterion (Edwards, 1989). Adjustment in the nominal exchange rate in order to avoid appreciation in the real rate, could, however, help avoid a balance of payments deficit. Bird and Orme (1981) also found it difficult to explain the apparent reluctance of countries to borrow from the IMF in terms of conditionality although Cornelius (1987b) claims that changes in conditionality over time do have a statistically significant effect on borrowing.

62 Support for this is provided in the main text where it is shown that countries often have a prolonged involvement with the Fund spanning many years.

63 For an interesting analysis of the politics of IMF conditionality, which discusses the causes of differential treatment, see Stiles (1990).

64 Bird and Orme (1981) found that on the basis of their models only 8 out of 27 non-drawing countries had predicted drawings of zero.

65 Eichengreen (1993) correctly points out that in principle the availability of IMF credit can have the same effect on government behaviour as deposit insurance may have on a bank, in that it encourages risk-taking by offering a bail-out. He suggests that the evidence that during the Bretton Woods era developing countries which ran especially expansionary policies and suffered large deteriorations in their balance of payments were more inclined to turn to the Fund (Edwards and Santaella, 1993) is consistent with the moral hazard view. The problem is that it is also consistent with other views. One might expect countries that turn to the Fund to have relatively severe economic problems. Overall the evidence presented here is that countries in general endeavour to avoid turning to the Fund even when their economic situation is weak; borrowing from the Fund is not perceived as a soft option. Moreover, the evidence of a connection between over-expansionary domestic policies and Fund borrowing was stronger during the Bretton Woods era, upon which Eichengreen was focusing, than subsequently. More recent evidence suggests that exogenous factors, even including natural disasters, may be a frequent proximate cause of Fund borrowing and it is difficult to see how the moral hazard view applies to these (Killick *et al.*, 1992).

66 It is interesting to note, however, that the adoption of Fund-type programmes in Latin America in the late 1980s and early 1990s coincided with the apparent restoration of creditworthiness. Yet it may be that this had more to do with relatively low interest rates in the United States than rising confidence in Latin America's long-term future (Bird and Helwege, 1994).

67 The optimum adjustment path for developing countries will not only be constrained by the availability of financing but also by the macro-economic and commercial policies pursued by industrial economies. An enduring problem for the Fund is how to put pressure on these countries to pursue policies that will facilitate adjustment in other parts of the world when they themselves are not looking for resources from the Fund.

68 There is evidence that the Fund is moving in this direction, although elements of change are superimposed on a strong basis of continuity in the design of Fund-backed programmes; see Killick (1992) for an assessment of the evidence.

69 One could envisage a conditionality Laffer Curve in which enhanced conditionality is initially associated with improving economic performance, even though at some point diminishing returns set in. Beyond a maximum, however, further conditionality so reduces commitment that performance deteriorates. If a country is located to the right of the maximum point on the conditionality Laffer Curve it follows that reduced conditionality will increase the commitment to adjust to such an extent that economic performance improves. In this sense it is possible to conceive of optimal, sub-optimal, and excessive conditionality. Maximising the catalytic effect of Fund-backed programmes therefore involves identifying optimal conditionality.

70 This trend has recently been illustrated by the introduction of a new temporary lending facility (the Systemic Transformation Facility) for those economies in Eastern and Central Europe which are in transition from centrally planned to more market-orientated systems.

71 Bird (1993c) provides a more detailed analysis of the case for resurrecting a form of link. The case is particularly compelling for low-income countries.

72 The real resource cost may be only temporary since countries supplying exports will accumulate additional reserves which represent a future claim on real resources. In such circumstances even a short-run real resource cost may imply zero welfare cost. Moreover, to the extent that additional export demand converts into additional aggregate supply the real resource cost may itself be transient.

# REFERENCES

Bergsten, C. Fred (1990) 'From Cold War to Trade War', *International Economic Insights*, July/August.

Bienen, H.S. and Gersovitz, M. (1986) 'Economic Stabilisation, Conditionality and Political Stability', *International Organisation*, Autumn.

Bird, Graham (1978) *The International Monetary System and the Less Developed Countries*, London: Macmillan.

——— (1979) 'The Benefits of Special Drawing Rights for Less Developed Countries', *World Development*, March.

——— (1981a) 'Reserve Currency Consolidation, Gold Policy and Financial Flows to Developing Countries: Mechanisms for an Aid-augmented Substitution Account', *World Development*, July.

——— (1981b) 'SDR Distribution, Interest Rates and Aid Flows', *The World Economy*, December.

——— (1982) 'Developing Country Interests in Proposals for International Monetary Reform', in Tony Killick (ed.), *Adjustment and Financing in the Developing World*, Washington, DC: IMF and London: Overseas Development Institute.

——— (1983) 'Low Income Countries and International Financial Reform', *Journal of Developing Areas*, October, reprinted in G. Bird, *Managing Global Money*, London: Macmillan.

——— (1985) *World Finance and Adjustment: An Agenda for Reform*, London: Macmillan.

——— (1987) *International Financial Policy and Economic Development: A Disaggregated Approach*, London: Macmillan.

——— (1988a) *Managing Global Money*, London: Macmillan.

——— (1988b) 'The Mix Between Adjustment and Financing', in G. Bird, *Managing Global Money*, London: Macmillan.

——— (1989) *Commercial Bank Lending and Third World Debt*, London: Macmillan.

——— (1990a) 'The International Financial Regime and the Developing World', in Graham Bird (ed.), *The International Financial Regime*, London: Academic Press with Surrey University Press.

——— (ed.) (1990b) *The International Financial Regime*. London: Surrey University Press with Academic Press.

## REFERENCES

————— (1990c) 'Evaluating the Effects of IMF-Supported Adjustment Programmes: An Analytical Commentary on Interpreting the Empirical Evidence', in K. Phylaktis and M. Pradhan (eds), *International Finance and Less Developed Countries*, London: Macmillan.

————— (1992a) 'Global Environmental Degradation and International Resource Transfers', *Global Environmental Change*, September.

————— (1992b) *IMF Lending: the Empirical Evidence*, ODI Working Paper No. 70, London: Overseas Development Institute.

————— (1993a) 'Sisters in Economic Development: The Bretton Woods Institutions and Developing Countries', *Journal of International Development*, Vol. 5, No. 1.

————— (1993b) 'Changing Partners and the Changing Perspectives and Policies of the Bretton Woods Institutions', mimeo, paper presented to the UN North–South Roundtable, New York, April.

————— (1993c) 'Economic Assistance to Low Income Countries: Should the Link Be Resurrected?', Surrey Centre for International Economic Studies Working Paper No. 5, September.

Bird, Graham and Bedford, David (1992) 'Why Do Countries Borrow from the IMF? An Analysis of the Empirical Evidence', Surrey Centre for International Economic Studies Working Paper No. 93/3, University of Surrey.

Bird, Graham and Helwege, Ann (eds) (1994) *Latin America's Economic Future?*, London: Academic Press.

Bird, Graham and Orme, Timothy (1981) 'An Analysis of Drawings on the IMF by Developing Countries', *World Development*, June.

Brandt, W. (1980) *North–South: A Programme for Survival*, London: Pan.

Brodsky, D.A. and Sampson, G.P. (1981) 'Implications of the Effective Revaluation of Reserve Asset Gold: The Case for a Gold Account for Development', *World Development*, July.

Cassen, R. and Associates (1987) *Does Aid Work?*, Oxford: Oxford University Press.

Chrystal, K.A. (1978) *International Money and the Future of the SDR*, Essays in International Finance, Princeton, NJ: Princeton University Press, June.

————— (1990) 'International Reserves and International Liquidity: A Solution in Search of a Problem', in G. Bird (ed.), *The International Financial Regime*, London: Academic Press with Surrey University Press.

Clark, P.B. (1970) 'Optimum International Reserves and the Speed of Adjustment', *Journal of Political Economy*, March/April.

Coats, Warren L. (1990) 'Enhancing the Attractiveness of the SDR', *World Development* 18(7): 975–88.

Coats, Warren L., Furstenberg, R.W. and Isard, P. (1990) *The SDR System and the Issue of Resource Transfers*, Essays in International Finance No. 180, Princeton, NJ: Princeton University Press, December.

Conway, Patrick (1991) 'IMF Lending Programmes: Participation and Impact', mimeo., April.

Corden, Max (1988) 'An International Debt Facility', *IMF Staff Papers*, June.

Cornelius, Peter (1987a) 'The Demand for IMF Credits by Sub-Saharan African Countries', *Economics Letters* 23.

———— (1987b) 'A Combined Cross-Section/Time Series Analysis of the Demand for IMF Credit by the Non-Oil Developing Nations', *Kredit und Kapital*, Vol. 1.

———— (1987c) 'On the Variation of Conditionality, Associated with IMF-Supported Adjustment Programmes', *Schweitzer Zeitschrift für Volkswirtschaft und Statistik*, June.

Cornia, G.A., Jolly, R. and Stewart, F. (eds) (1987) *Adjustment with a Human Face: Protecting the Vulnerable and Promoting Growth*, Oxford: Oxford University Press.

Dell, S. (1982) 'Stabilization: The Political Economy of Overkill', *World Development*, August.

———— (1985) 'The Fifth Credit Tranche', *World Development*, February.

Dell, S. and Lawrence, R. (1980) *The Balance of Payments Adjustment Process in Developing Countries*, New York: Pergamon.

Demery, Lionel and Addison, Tony (1987) *The Alleviation of Poverty Under Structural Adjustment*, Washington, DC: World Bank.

Edwards, S. (1983) 'The Demand for International Reserves and Exchange Rate Adjustments: the Case of LDCs 1964–72', *Economica*, August.

———— (1988) *Exchange Rate Misalignment in Developing Countries*, Baltimore, MD: Johns Hopkins University Press.

———— (1989) 'The International Monetary Fund and the Developing Countries: A Critical Evaluation', *Carnegie Rochester Conference Series on Public Policy* 31.

Edwards, S. and Montiel, P. (1989) 'Devaluation Crises and Macroeconomic Consequences of Postponed Adjustment in Developing Countries', *IMF Staff Papers*, December.

Edwards, S. and Santaella, Julio A. (1993) 'Devaluation Controversies in the Developing Countries: Lessons from the Bretton Woods Era', in Michael D. Bordo and Barry Eichengreen (eds), *A Retrospective on the Bretton Woods System: Lessons for International Monetary Reform*, Chicago, IL: University of Chicago Press.

Eichengreen, Barry (1993) 'Epilogue: Three Perspectives on the Bretton Woods System', in Michael D Bordo and Barry Eichengreen (eds), *A Retrospective on the Bretton Woods System: Lessons for International Monetary Reform*, Chicago, IL: University of Chicago Press.

Feinberg, Richard E. (1988) 'The Changing Relationship Between the World Bank and the International Monetary Fund', *International Organization*, Summer.

Finch, David C. (1988) 'Let the IMF be the IMF', *International Economy*, January/February.

———— (1989) *The IMF: The Record and the Prospect*, Essays in International Finance No. 175, Princeton, NJ: Princeton University Press, September.

Frenkel, J.A. (1984) 'International Liquidity and Monetary Control', in G.M. von Furstenberg (ed.), *International Money and Credit: The Policy Roles*, Washington, DC: IMF.

## REFERENCES

Goreux, Louis M. (1989) 'The Fund and the Low Income Countries', in Catherine Gwin and Richard E. Feinberg (eds), *The International Monetary Fund in a Multipolar World: Pulling Together*, US–Third World Policy Perspectives No. 13, Washington, DC: Overseas Development Council.

Griffith-Jones, Stephany and Lipton, Michael (1984) 'International Lenders of Last Resort: Are Changes Required?', Occasional Papers in International Trade and Finance, Midland Bank, March.

Guitian, M. (1991a) 'What Have We Learned from the International Debt Crisis', mimeo, March.

—— (1991b) 'The Process of Adjustment and Economic Reform: Real and Apparent Differences Between East and West', mimeo, April.

—— (1991c) 'The Route Toward a Single European Market: The Need for More than One Track', mimeo, April.

Gwin, Catherine and Feinberg, Richard (eds) (1989) *The International Monetary Fund in a Multipolar World: Pulling Together*, US–Third World Policy Perspectives No. 13, Washington, DC: Overseas Development Council.

Haggard, Stephan (1985) 'The Politics of Adjustment: Lessons from the IMF's Extended Fund Facility', *International Organisation*, 39(s).

Haggard, Stephan and Kaufman, Robert (1989) 'The Politics of Stabilization and Structural Adjustment', in Jeffrey D. Sachs (ed.), *Developing Country Debt and Economic Performance, Vol. 1. International Financial System*, Chicago, IL: University of Chicago Press.

Helleiner, G.K. (1983a) 'Lender of Early Resort: The IMF and the Poorest', *American Economic Review*, May.

—— (1983b) *The IMF and Africa in the 1980s*, Essays in International Finance No. 152, Princeton, NJ: Princeton University Press, July.

—— (1986) 'Balance of Payments Experience and Growth Prospects of Developing Countries: A Synthesis', *World Development*, August.

—— (1987) 'Stabilization, Adjustment and the Poor', *World Development*, December.

Hewitt, Adrian, and Killick, Tony (1993) 'Bilateral Aid Conditionality: A First View', mimeo, London: Overseas Development Institute.

IMF (1986) *Fund-Supported Programmes, Fiscal Policy and Income Distribution*, IMF Occasional Paper No. 46.

Joyce, Joseph P. (1992) 'The Economic Characteristics of IMF Programme Countries', *Economics Letters* 38.

Kafka, Alexandre (1991) *Some IMF Problems after the Committee of Twenty*, paper prepared for a conference in honour of Jacques J. Polak, mimeo.

Kaufman, Robert R. (1988) *The Politics of Debt in Argentina, Brazil and Mexico*, Berkeley, CA: Institute of International Studies, University of California.

Kenen, Peter B. (1985) *Financing, Adjustment and the IMF*, Studies in International Economics, Washington, DC: Brookings Institution.

Khan, Mohsin S. (1990) 'The Macroeconomic Effects of Fund-Supported Adjustment Programmes', *IMF Staff Papers*, June.

—— (1991) 'Evaluating the Effects of IMF-Supported Adjustment Programmes: An Empirical Survey', in K. Phylaktis and M. Pradhan (eds), *International Finance and Less Developed Countries*, London: Macmillan.

# REFERENCES

Khan, M.S. and Knight, M.D. (1983) 'Determinants of Current Account Balances of Non-Oil Developing Countries in the 1970s: An Approach Empirical Analysis', *IMF Staff Papers*, December.

———— (1988) 'Import Compression and Export Performance in Developing Countries', *Review of Economics and Statistics*, May.

Khan, Mohsin, Montiel, Peter and Ul Haque, Nadeem (1986) 'Adjustment with Growth: Relating the Analytical Approaches of the World Bank and the IMF', *World Bank Discussion Paper,* Washington, DC: World Bank.

Killick, Tony (1989a) *A Reaction Too Far: Economic Theory and the Role of the State in Developing Countries,* London: Overseas Development Institute.

———— (1989b) 'Issues Arising from the Spread of Obligatory Adjustment', in Graham Bird (ed.), *Third World Debt: The Search for a Solution,* London: Edward Elgar.

———— (1992) *Continuity and Change in IMF Programme Design, 1982–92,* ODI Working Paper 69, December, London: Overseas Development Institute.

Killick, Tony, Bird, Graham, Sharpley, Jennifer and Sutton, Mary (1984) *The Quest for Economic Stabilisation: The IMF and the Third World,* London: Heinemann in association with the Overseas Development Institute.

Killick, Tony, Malik, Moazzam and Manuel, Marcus (1992) 'What Can We Know About the Effects of IMF Programmes?', *The World Economy,* September.

Krugman, Paul (1988) 'Financing versus Forgiving a Debt Overhang', *Journal of Development Economics* 29.

Lal, D. (1980) *A Liberal International Economic Order: The International Monetary System and Economic Development,* Essays in International Finance, Princeton, NJ: Princeton University Press, October.

Lessard, D. and Williamson, J. (1985) *Financial Intermediation Beyond the Debt Crisis,* Washington, DC: Institute for International Economics.

Llewellyn, D.T. (1990) 'The International Capital Transfer Mechanism of the 1970s: A Critique', in G. Bird (ed.), *The International Financial Regime,* London: Surrey University Press with Academic Press.

Love, James (1990) 'Export Earnings Instability: The Decline Reversed?', *Journal of Development Studies,* June.

Loxley, John (1984) *The IMF and the Poorest Countries,* Ottawa: North–South Institute.

Mason, Edward S. and Asher, Robert E. (1973) *The World Bank Since Bretton Woods,* Washington, DC: Brookings Institution.

Maynard, Geoffrey and van Ryckeghem, Willy (1976) *A World of Inflation,* London: Batsford.

Moggridge, Donald (ed.) (1980) *The Collected Writings of John Maynard Keynes, Vol. XXVI, Activities 1941–46,* London: Macmillan.

Nelson, Joan M. (1984) 'The Politics of Stabilisation', in R.E. Feinberg and V. Kallab (eds), *Adjustment Crisis in the Third World,* Washington, DC: Overseas Development Council.

Nelson, Joan M. (ed.) (1989) *Fragile Coalitions: The Politics of Economic Adjustment,* US–Third World Policy Perspectives No. 12, Washington, DC: Overseas Development Council.

# REFERENCES

Nowzad, B. (1981) *The IMF and Its Critics*, Essays in International Finance No. 146, Princeton, NJ: Princeton University Press, December.

——— (1989) 'The Debt Problem and the IMF's Perspective', in Graham Bird (ed.), *Third World Debt: The Search for a Solution*, London: Edward Elgar.

Otani, Ichiro and Villanueva, D. (1990) 'Long-term Growth in Developing Countries and its Determinants: An empirical analysis', *World Development* 18(6): 769–83.

Polak, Jacques J. (1991) *The Changing Nature of IMF Conditionality*, Essays in International Finance No. 184, Princeton, NJ: Princeton University Press, September.

Sachs, Jeffrey D. (1989a) 'Strengthening IMF Programmes in Highly Indebted Countries', in C. Gwin and R. Feinberg (eds), *The International Monetary Fund in a Multipolar World: Pulling Together*, US–Third World Policy Perspectives No. 13, Washington, DC: Overseas Development Council.

——— (1989b) 'Conditionality, Debt Relief, and the Developing Country Debt Crisis', in Jeffrey D. Sachs (ed.), *Developing Country Debt and Economic Performance, Vol. 1. International Financial System*, Chicago, IL: University of Chicago Press.

Sidell, Scott R. (1988) *The IMF and Third World Instability: Is There a Connection?*, London: Macmillan.

Stallings, Barbara and Kaufman, Robert (1989) *Debt and Democracy in Latin America*, Boulder, CO: Westview Press.

Stiles, Kendall W. (1990) 'IMF Conditionality: Coercion or Compromise?' *World Development*, July.

Vaubel, R. (1983) 'The Moral Hazard of IMF Lending', *The World Economy*, September.

——— (1986) 'A Public Choice Approach to International Organisation', *Public Choice* 51.

——— (1991) 'The Political Economy of the International Monetary Fund: A Public Choice Analysis', in R. Vaubel and T.D. Willett (eds), *The Political Economy of International Organizations*, Boulder, CO: Westview Press.

Vines, David (1990) 'Growth Oriented Adjustment Programmes; A Reconsideration', London: Centre for Economic Policy Research Discussion Paper No. 406, March.

Von Furstenberg, G.M. (ed.) (1984) *International Money and Credit: The Policy Roles*, Washington, DC: IMF.

Williamson, J. (1983) 'The Lending Policies of the International Monetary Fund', in J. Williamson (ed.), *IMF Conditionality*, Washington, DC: Institute for International Economics.

——— (1984) *A New SDR Allocation*, Washington, DC: Institute for International Economics.

——— (1992) 'International Monetary Reform and the Prospects for Economic Development', paper presented at a workshop organised by the Forum on Debt and Development, The Hague, mimeo, June.

Zulu, Justin B. and Nsouli, Saleh M. (1985) *Adjustment Programmes in Africa: The Recent Experience*, IMF Occasional Paper No. 34, April, Washington, DC: IMF.

# INDEX

Note: Abbreviations are used as in the text
Page numbers in **bold** type refer to **figures**
Page numbers in *italic* type refer to *tables*

185